54 Advances in Polymer Science

Fortschritte der Hochpolymeren-Forschung

Spectroscopy: NMR, Fluorescence, FT-IR

With Contributions by
C. W. Frank, J. L. Koenig, E. D. v. Meerwall,
S. N. Semerak

With 40 Figures and 5 Tables

Springer-Verlag Berlin Heidelberg GmbH 1984

ISBN 978-3-662-15249-2 ISBN 978-3-540-38655-1 (eBook)
DOI 10.1007/978-3-540-38655-1

Library of Congress Catalog Card Number 61-642

Originally published by Springer-Verlag Berlin Heidelberg New York Tokyo in 1984

Softcover reprint of the hardcover 1st edition 1984

2152/3020–543210

Editors

Editorial

With the publication of Vol. 51, the editors and the publisher would like to take this opportunity to thank authors and readers for their collaboration and their efforts to meet the scientific requirements of this series. We appreciate our authors concern for the progress of Polymer Science and we also welcome the advice and critical comments of our readers.

With the publication of Vol. 51 we should also like to refer to editorial policy: *this series publishes invited, critical review articles of new developments in all areas of Polymer Science in English (authors may naturally also include works of their own).* The responsible editor, that means the editor who has invited the article, discusses the scope of the review with the author on the basis of a tentative outline which the author is asked to provide. Author and editor are responsible for the scientific quality of the contribution; the editor's name appears at the end of it.
Manuscripts must be submitted, in content, language and form satisfactory, to Springer-Verlag. Figures and formulas should be reproducible. To meet readers' wishes, the publisher adds to each volume a "volume index" which approximately characterizes the content.

Editors and publisher make all efforts to publish the manuscripts as rapidly as possible, i.e., at the maximum, six months after the submission of an accepted paper. This means that contributions from diverse areas of Polymer Science must occasionally be united in one volume. In such cases a "volume index" cannot meet all expectations, but will nevertheless provide more information than a mere volume number.

From Vol. 51 on, each volume contains a subject index.

Editors | Publisher

Table of Contents

Self-Diffusion in Polymer Systems, Measured with Field-Gradient Spin Echo NMR Methods

Ernst D. von Meerwall
Physics Department, The University of Akron, Akron, Ohio 44325, U.S.A.

The steady gradient and pulsed gradient spin echo NMR Methods of measuring self-diffusion have for some twenty years been applied to the study of polymers. The methods are briefly described, and the principal results of this research are reviewed in three main areas: diffusion of polymers in the melt and in concentrated solutions, diffusion of polymer in dilute and semidilute solutions, and diffusion of penetrants and diluents in high polymers hosts. The theoretical interpretations of these experiments are included in the review, with particular attention to theories of dilute polymer solutions, the free-volume theory of diffusion in concentrated solutions, and power-law behavior postulated for various regimes.

The aim of this review is to familiarize workers in the polymer field with these techniques for measuring self-diffusion and with their applications and benefits.

List of Symbols and Abbreviations

A	height of spin echo in pulsed NMR, measured as function of field gradient parameters or time.
α	Solvent power parameter entering Flory's theory of dilute solutions.
α'	degree of neutralization in polyelectrolyte solutions.
B	free-volume parameter entering Vrentas-Duda theory; subscript (1,2,i) denotes molecular species in solution.
B_d	ratio of vacancy size needed for diluent diffusion to diluent size, in the Fujita-Doolittle theory.
c	polymer concentration in solutions (mass/solution volume).
c*, c**	lower and upper limits of c in semidilute regime to which scaling laws apply.
C_1, C_2	parameters entering the Williams-Landel-Ferry equation.
CM	referring to center-of-mass motion of a macromolecule.
γ	magnetogyric ratio of nucleus at resonance (frequency/field).
d	diffusion distance.
D	diffusion coefficient; subscript indicates species diffusing.
D_0, D_i^0	diffusion coefficient of a given species (i) in the limit of zero concentration in solution.
D_0, D_∞	diffusion coefficients measured at very short and very long diffusion times t.
D_{max}, D_{min}	diffusion coefficients measured at very small and very large values of gradient parameter δ^2G^2 at fixed diffusion time.
D_c	Cooperative diffusion coefficient.
δ	duration of magnetic field gradient pulse.
Δ	interval between (beginning of) magnetic field gradient pulses.
$\Delta\alpha$	thermal expansivity of free volume above the glass transition.
E_a	activation energy for self-diffusion.
η_0	viscosity of solvent.
f	fractional free volume; subscript (i, 1, 2, p, dil) denotes molecular species in solution.
f_g	fractional free volume at and below the glass transition.
F	functionality, number of arms of a star-branched polymer.
FGSE	Field gradient spin echo method of measuring self-diffusion, encompasses SGSE and PGSE variants.
FT	Fourier transform; refers to NMR experiments in which the time domain response of the spin system is transformed into a frequency spectrum.
g	overlap factor; describes multiple access to hole free volume for molecular transport.
G	magnitude of pulsed magnetic field gradient.
G_0	magnitude of steady magnetic field gradient.
k	Boltzmann constant
k_F	lowest-order coefficient of concentration dependence of 1/D in dilute polymer solutions.

λ	extension ratio.
M, M_c	molecular weight; at the onset of entanglements.
$\bar{M}_n$, $\bar{M}_w$	number average and weight average molecular weight of polydisperse polymer.
m	collision-dynamic mass entering the Vrentas-Duda theory; subscript indicates molecular species in solution.
n	exponent of molecular weight in scaling laws.
N	number of main-chain carbon atoms in polymer molecules.
NMR	nuclear magnetic resonance, pulsed or continuous-wave (CW).
PGSE	pulsed-gradient spin echo method of measuring self-diffusion.
rf	radio frequency; refers to NMR spectroscopy.
R	universal gas constant.
ϱ	density (mass/volume).
SGSE	steady-gradient spin-echo method of measuring self-diffusion.
$\langle s^2 \rangle$	mean squared linear dimension of coiled macromolecule in solution or in melt.
t	diffusion time.
t_1	time between radio frequency pulse and onset of gradient pulse (PGSE).
T	(absolute) temperature.
T_g, $T_{g\infty}$	glass transition temperature; at infinite molecular weight.
T_1, $T_{1\varrho}$	nuclear spin lattice relaxation time; in the rotating frame.
T_2	nuclear spin-spin relaxation time.
τ	time between 90° and 180° rf pulses in T_2 or FGSE experiment.
v	volume fraction; subscript (1, 2, i, p, dil) denotes species in solution.
V_E	molar free volume contributed by the ends of chain molecules.
V_F	molar free volume from any source.
$\hat{V}_i^*$	specific critical volume for diffusion of one molecule or segment of species i.
w_i	weight fraction of species i in solution.

1 Introduction

Nuclear magnetic resonance (NMR) is used in the study of polymers not only as a tool to characterize chemical structure, but also to measure molecular motion in aggregates of polymer molecules such as solutions, melts, and entangled or cross-linked networks [1,2]. This is done by observing either the width of the spectral features (CW NMR) or more directly (pulsed NMR) by measuring characteristic NMR relaxation times, the spin-spin relaxation time T_2 and the spin-lattice relaxation times T_1 or $T_{1\varrho}$. The molecular motions reflected in these quantities include vibrations and rotations of molecules or their articulated moieties, as well as spatial translations among molecules, e.g., diffusive motions of macromolecules, or of penetrant or diluent molecules within polymers. Under favorable conditions it is possible to partition the observed relaxation rates T_1^{-1} or T_2^{-1} to obtain, somewhat indirectly, a measure of each contribution, including those from diffusive motion. It is generally necessary to measure the relaxation times over a range of temperatures to obtain this partitioning.

It is, however, possible to use pulsed-NMR spectrometers to measure diffusive translational motion of molecules directly. This ability depends on taking control of the cause of the translational contribution to T_2^{-1}. Early in the history of NMR it was shown [3,4] that it is possible to produce a "spin echo", a refocusing of nuclear magnetization which appears at time τ after the second rf pulse in a two-pulse sequence 90°-τ-180°. The angles refer to the rotations of the nuclear magnetization vector from its original direction along the magnetic field. This pulse sequence (or variations of it) is used to measure T_2; it was soon observed that in rapidly diffusing specimens the measured "T_2" became shortened in proportion to the inhomogeneities in the magnetic field in the region of the sample [3,4]. This effect, once analyzed, [4] became the basis of all field-gradient spin-echo (FGSE) methods of measuring self-diffusion: a set of measurements of the magnitude of the spin echo as function of the magnitude and duration of the calibrated field gradient yields the diffusion constant D of the species at resonance. This article will confine itself to reviewing diffusion measurements in polymer systems made with these methods, the steady gradient spin echo (SGSE) method and its pulsed-gradient (PGSE) variant [5,6].

Because field-gradient spin-echo measurements of D depend on no driving force such as a concentration, temperature, or velocity gradient, etc., they reflect Brownian motion of the molecules in the laboratory reference frame, and are usually referred to as self-diffusion. These attributes are further discussed in Sections 2 and 3.

In polymers, the field-gradient spin-echo methods of measuring self-diffusion have been useful in three more or less distinct areas, the diffusion of polymers in their own melt and in concentrated solutions, in dilute and semidilute solutions, and the diffusion of penetrants and diluents in polymer hosts. A fourth category, the diffusion of bulky or flexible molecules in polymer hosts, is useful for subject matter not closely associated with the first and third category. It should be noted that the work reviewed here represents only a small fraction of the diffusion studies in polymers, including those using other NMR methods.

2 Techniques

2.1 Equipment

SGSE and PGSE diffusion measurements require a pulsed NMR spectrometer with a provision for creating a uniform calibrated magnetic field gradient in the region of the sample [4,5,6]. The spectrometer may operate in the conventional or in the Fourier Transform (FT) mode. The typical pulse sequence is the 90°-τ-180° echo sequence, particularly for SGSE work, but others, such as the Carr-Purcell sequence ("method B") [4], are used as well [7]. Particular pulse sequences elicit a spin echo at long times in highly viscous samples [8] or permit diffusion measurements under difficult experimental conditions [9].

The magnetic field gradient is usually generated by an electric current through a set of coils centered on the sample. These coils may be of the opposed Helmholtz design [6] or else of the quadrupole type [10]. The latter's more compact construction often permits their incorporation into the NMR probe assembly, and thus minimizes problems associated with eddy current and magnetic image effects [6,10] in PGSE work using iron-core magnets. PGSE work requires fast response of the coil current power supply and switch, combined with precise current regulation and precise timing of the current pulses. Time and phase stability of the spin echo also depends on physical rigidity of the gradient coils and probe assembly. Advances in these areas [11] have eliminated many of the early difficulties experienced by PGSE workers with respect to stability and reproducibility of the spin echo height, particularly at lower values of diffusion coefficient. The magnetic field gradient may be calibrated by a direct mapping the magnetic field [11], by a first-principles calculation [10] (including the effects of eddy currents and magnetic images in the case of PGSE), by relating the duration of the spin echo for a given sample diameter to the coil current [4], e.g. from measurements of the spin echo envelope during off-resonance FT PGSE experiments [12], or ultimately by echo attenuation measurements in substances of known diffusivity. The problems associated with short and long-term residual gradients, such as the rf phase shifts encountered in non-FT PGSE work using phase-sensitive detection, have now been analyzed [12]. Many spin-echo spectrometers are computerized [13,14] to permit data taking and signal averaging as well as performing some prelimiary analysis; the most advanced equipment [11,13] also permits computer control of rf pulse sequences and gradient pulses (often through recourse to the coils intended to trim the magnetic field or manipulate its homogeneity [15]). Essentially fully automatic operation of SGSE work has been achieved [13], and is nearly attained [11,16] for routine PGSE work.

2.2 Experiments

The parameter to be measured in spin-echo diffusion experiments is the amplitude A of the echo. Its magnitude depends [3] on the spin-spin relaxation time T_2 and on the diffusion coefficient D. For an SGSE experiment employing a 90°-τ-180° rf pulse sequence in the presence of a steady field gradient of magnitude G_0 [4],

$$A(2\tau) = A(0) \exp -\left[\frac{2\tau}{T_2} + \frac{2\gamma^2 G_0^2 (2\tau)^3}{3} D\right] \tag{1}$$

where γ is the magnetogyric ratio of the nucleus at resonance. Because the transverse magnetization decay may not be strictly exponential and T_2 may not be precisely known, it is customary to avoid dealing with the first term in the exponential by conducting SGSE experiments at fixed τ, varying G_0. A plot of log $A(2\tau,G_0)/(2\tau, G_0 = 0)$ *vs.* G_0^2 then yields a straight line whose slope is proportional to D. If two or more molecular species are simultaneously at resonance, with distinct T_2 and D values, the single exponential in eq. (1) must be replaced by a weighted sum, the weights representing the relative numbers of nuclei per species [17,18].

Since the spin-echo time duration is inversely proportional [3,4] to G_0 (an effect independent of diffusivity), there are practical limits to the magnitude of G_0: the echo cannot be detected as it becomes shorter than the inverse audio band width of the NMR receiver. For this reason, the lowest values of D measurable with SGSE techniques are on the order of $D \approx 5 \times 10^{-8}$ cm²/sec. Another two orders of magnitude may be gained by arranging for the field gradient to have a small magnitude G_0 during the rf pulses and the echo, and a much larger strength $G + G_0$ for duration δ after the 90° pulse and again after the 180° pulse [5,6] (see Fig. 1). Provided that these gradient pulses are accurately balanced ($G \cdot \delta$ identical) the echo will again occur at $t = 2\tau$ and have the relative magnitude [5,6]

$$\frac{A(2\tau, G, G_0)}{A(2\tau, G = 0, G_0)} = \exp(-\gamma^2 Dx) \tag{2}$$

where $x = \delta^2 G^2(\Delta - \delta/3)$

$$-\delta G G_0[(t_1^2 + t_2^2) + \delta(t_1 + t_2) + 2\delta^2/3 - 2\tau^2],$$

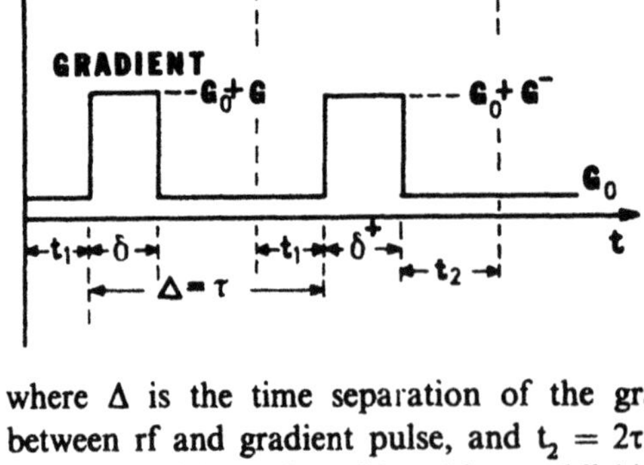

Fig. 1. Radio frequency (top) and magnetic field gradient (bottom) pulse sequence for the basic PGSE experiment (Ref. [14], by permission). Advances in coil current regulation have obviated the necessity of adjusting δ to keep $\delta \cdot G$ constant.

where Δ is the time separation of the gradient pulses, t_1 represents the delay between rf and gradient pulse, and $t_2 = 2\tau - \Delta - \delta - t_1$. G_0 and G are usually parallel, which can be achieved by establishing a current leak through the switching circuit which provides the pulses producing G. G_0 is usually kept non-zero (but $G_0 \ll G$) both to narrow the echo sufficiently to give access to the signal baseline for measuring A, and to increase the stability of the echo [12]. Ignoring the second term in x may introduce offset or curvature into the plot of log A vs. x and falsify D.

Actual curvature in such a plot may signal the presence of a distribution of diffusivities [6], as in polymer polydispersity. This hypothesis may be modelled and compared with experiment [19].

PGSE experiments are usually conducted at fixed Δ(or τ), varying δ and/or G. In that case the diffusion time is relatively well-defined and nearly constant [5,6]:

$$t = \Delta - \delta/3 \qquad (\delta \ll \Delta)\,, \tag{3}$$

since most of the echo attenuation is usually due to G rather than G_0. Thus, PGSE offers an opportunity of testing explicitly for a time-dependence of D (non-Fickian diffusion) by measuring A vs. Δ. The practical limits on 2τ (hence Δ, since $\Delta < 2\tau$) are usually about 1 ms $< 2\tau <$ 500 ms in polymer systems. Thus, a curvature in the plot of log A vs. x (varying Δ and perhaps τ, in the absence of polydispersity) signifies restricted diffusion or some other mechanism at variance with the random-walk nature of molecular motion. From the initial slope-$\gamma^2 D_0$, final slope-$\gamma^2 D_\infty$, and the time t′ of crossover between initial and final slopes one may determine not only the short-range diffusivity D_0 but also the distance scale of the short-range motions (e.g. barrier spacing) and the impediments encountered in long-range motion (barrier permeability, etc.). Appropriate mathematical expressions [6,20] (many suitable for curve-fitting [21,22]) for this and many other diffusion hypotheses are found in the literature; computer programs for data interpretation are also available [23,24]. These are useful whenever diffusion is microscopically inhomogeneous or anisotropic; it is precisely in such situations that PGSE (and often SGSE) techniques enjoy the greatest advantages over other methods of measuring diffusion. In polymer systems, these capabilities have had only limited applications; obvious opportunities are diffusion of diluents in filled or partially crystalline polymers.

Other types of PGSE experiments, including those involving alternating or sinusoidal field gradient pulses [25], have particular applications which may be useful in polymer systems.

3 Capabilities and Limitations

3.1 General

Spin-echo diffusion measurements measure the Brownian motion of all nuclei at resonance, hence, of the molecules of which they are part. This is usually referred to as self-diffusion, no matter how many molecular species are simultaneously present or diffusing. The label employed to track the nuclei is the direction of their magnetic moments; the variable used to map position is the strength of the applied magnetic field, which varies linearly across the sample in a specified direction along which diffusion is measured. The reference frame is laboratory-fixed, with no net mass transport during the experiment. The experiments operate without concentration or temperature gradients; a single sample (at most a few cm^3 of material) may be reused indefinitely, e.g., to study temperature-dependence. No tracers or radioactive materials are used; the sample is not physically agitated or altered and thus can remain sealed within a sample tube during all measurements.

3.2 Experimental uncertainty

Precision in the measured diffusivities is limited by the reproducibility of the echo height measurements and by field gradient calibration. Under favorable circumstances, both of these can be kept below about 1%, although for less time-consuming and routine measurements, 2–3% random uncertainty is more typical. If comparisons among different samples are more important than comparisons with other work on similar samples, then calibration error is quasisystematic and secondary. In that case, perhaps 4% calibration uncertainty may be acceptable, requiring measurement of G or G_0 to within 2% (see Eqs. 1, 2).

The reproducibility of spin-echo measurements is a sensitive function of the temperature stability of the sample during measurement. Changes in temperature affect both the unattenuated echo height (mainly via T_2) and the degree of attenuation (via D).

3.3 Range, Resolution, Sensitivity

Diffusion coefficients measurable with SGSE and PGSE techniques encounter no upper limit likely to be reached in polymer systems. The lower limit for SGSE [18] is approx. $D_{min} \approx 2 \times 10^{-8}$ cm²/sec, for PGSE [6] $D_{min} \approx 10^{-10}$ cm²/sec. These limits are approached only in favorable cases; more typically they are higher by a factor of about 3–10, although polymer systems tend to be rather well suited for spin-echo work. This relatively high lower limit makes diffusion measurements in melts and concentrated solutions of high-molecular-weight polymers extremely difficult, and has so far prevented PGSE from addressing directly the problem of reptation [26,27] in the melt.

The lower limit of D in PGSE work is approached rather suddenly: as molecular motion decreases (at lower temperature or higher molecular weight, etc.), both T_2 and D tend to decrease. This confronts the experimenter with the impossibility of finding a value of τ small enough to result in a spin-echo of acceptable signal-to-noise ratio, but large enough to permit significant attenuation in the given time $\delta < \tau$ with the largest field gradients available. Near D_{min} random uncertainties in D may reach 20–30%.

^{1}H NMR is generally used for diffusion measurements in polymers since protons tend to be abundant and offer large NMR signal strength. Concentrations approaching 0.1 wt.% polymer in a proton-free solvent can be usefully studied using signal averaging, and less than 1 wt.% diluent can be observed to diffuse in rubber, in the presence of the rubber's own proton spin echo. Discrimination among different simultaneously diffusing species is accomplished in various ways. FT spectrometers can rely on differences in chemical shift for a simultaneous display of echo attenuation for each species [15,28]; this method fails when chemical shifts are closely similar, e.g. oligomers dissolved in high polymers of the same species. Differences in spin-lattice relaxation time T_1 may be used to "tune out" the echo of a species not under consideration [29]. Species may be distinguished merely from differences in their diffusivity; curve-fitting can resolve D values whose ratio exceeds 2 to 3 except where the ratio of echo heights arising from the two species is extreme [24,29]. A special example of this method is the subtraction of the spin echo contributed by an essentially non-diffusing species [30], e.g. a host polymer. Finally, use of mutually exclusive nuclides at resonance (e.g. ^1H, ^{19}F, ^{13}C) permits separate recording of the echo attenuation pertinent to a given molecular species; a separate experiment is necessary for each species in a solution [31].

Because T_2 tends to be longer for molecules of lower molecular weight, spin-echo experiments tend to be very sensitive to contamination by light, diffusing impurities often found in polymer preparations. Care must be taken to remove these [32] to avoid inviting misinterpretation of experimental data.

3.4 Distances Scales

For unrestricted Brownian motion of molecules the random-walk distance d is given by [6]

$$d = (2Dt)^{1/2}\,, \tag{4}$$

where for PGSE experiments t is given by Eq. (3). If d is substantially larger than the coiled dimension $\langle s^2\rangle^{1/2}$ of the macromolecule being observed, the experiment will record true center-of-mass motion, otherwise the measured diffusion coefficient will contain (probably non-Fickian) contributions from segmental mechanisms, incl. cooperative rotations and gel-like motions [33]. Since in polymer molecules of molecular weight M for most architectures $\langle s^2\rangle$ is proportional to M whereas D is a strongly decreasing function of M, both terms in the inequality

$$2Dt \gg \langle s^2\rangle \quad \text{(CM motion)} \tag{5}$$

will rapidly approach each other with increasing M. For typical polymers and experimental conditions this tends to occur near the lower limit of D measurable via PGSE. The departure from a strictly Fickian diffusivity should in principle be observable by a direct measurement of echo attenuation vs. Δ. However, near the lower limits accessible to PGSE work, little flexibility remains [6] for varying τ (see above) or Δ (since $\Delta \to \tau$ as $\delta \to \tau$), so that such a test has so far not been practicable in polymer melts or concentrated solutions. The situation in semidilute solutions will be discussed below.

4 Polymers in the Melt and in Concentrated Solutions

The first direct spin-echo diffusion measurement in a polymer melt was reported by McCall, Douglass, and Anderson [34] in low-molecular weight polyethylenes. Because of the limitations of the SGSE method the authors found themselves restricted ($D > 10^{-7}$ cm^2/sec) to molecular weights $M < 10^4$ and to temperatures T above 130 °C. They observed an Arrhenius behavior:

$$D(T) = D' \exp(-E_a/RT) \tag{6}$$

with an activation energy E_a which increased slightly with increasing M, whereas D itself at constant temperature T decreased with increasing M. These results represented a continuation of earlier work on n-paraffin hydrocarbons by Douglass and McCall [35], in which similar results were obtained for n-pentane through n-octadecane. The Eyring theory [36] for liquid diffusion partially explained their data. These two investigations together suggested that D at a given T is related to the molecular weight M by a power law:

$$D(M, T) = K(T)\, M^n \,, \tag{7}$$

with $n = -1.66$ at $T = 150$ °C and 200 °C. This exponent may be compared with the value -2.0 to -2.25 obtained in n-paraffins (n = 8 through 36) at 80 °C by von Meerwall and Ferguson [37]. The difference in exponents is probably related to the contribution to the free volume of the melt by the macromolecular chain ends, to be discussed below.

A detailed SGSE investigation of linear dimethylsiloxanes at lower M was conducted by McCall, Anderson and Huggins [38]. The dependence of D on molecular size was less than in n-paraffins, leading to the conclusion that diffusion in siloxanes is controlled to a much larger extent by chain configurational effects. Measurement of D as function of pressure showed a linear decrease of log D with increasing pressure, the slope being slightly steeper at high molecular weight.

Later, McCall and Huggins [18] extended this work to molecular weights of 15000 and 32000, the highest molecular weight polymers to be investigated with the SGSE method, and attained (with difficulty) diffusivities as low as $D = 2 \times 10^{-8}$ cm²/sec. The dependence of D at constant T on molecular weight followed the form of Eq. (7), with an exponent of -1.55, the monomeric friction coefficient increasing only slightly with molecular weight. The authors also analyzed the effect of polydispersity on SGSE results, demonstrating that misinterpretation of the data is possible if significant polydispersity is ignored.

In polyisobutylene in the melt and in solution (CCl_4, CS_2), McCall, Douglass, and Anderson [17] found that the activation energies for polymer diffusion increased with polymer concentration from the value at infinite dilution (approaching the pure solvent value) to the value in the melt. Solvent diffusion, and solvent effect on polymer diffusion, were also measured. The Stokes-Einstein model applied to this data yielded molecular dimensions too small by a factor of two or three.

Tanner's investigation [39] of the benzene-polydimethylsiloxane and chloroform-poly(ethylene oxide) systems employed the PGSE method and obtained diffusivities down to well below 10^{-9} cm²/sec, in polymers of molecular weights up to 800,000. He showed that the diffusion of the solvent, or of a low-molecular weight polymer fraction, depends on the molecular weight of the host polymer melt until the latter becomes significantly larger than the molecular weight of the smaller molecules. This finding was corroborated by viscosity measurements, and was qualitatively explained in hydrodynamic terms. Some of this data is shown in Fig. 2.

The concentration dependence of polymer diffusion of the same polymers in various solvents was explored by Tanner, Liu, and Anderson [40]. They were the first to observe via PSGE the sharp decrease of log D with increasing polymer concentration; the decrease was nearly linear at low molecular weights but became initially steeper

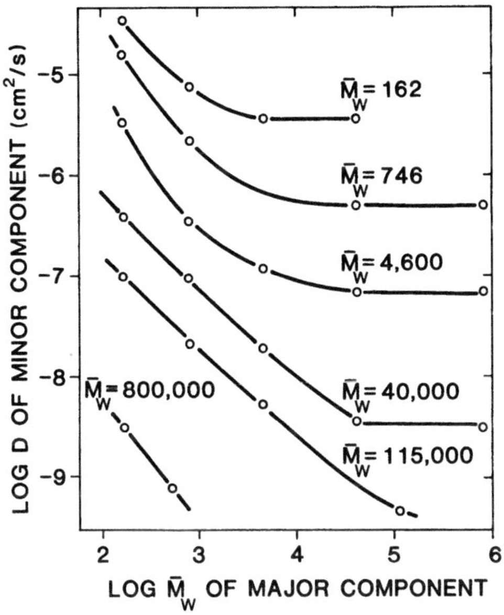

Fig. 2. Self-diffusion of component A in a binary solution of polydimethylsiloxanes (10% A, 90% B) as function of the molecular weight of B. Each curve represents a separate value of the molecular weight of A, labelled. (after Ref. [39], with permission).

and concave upward at higher molecular weights. Part of the upward concavity was tentatively attributed to the modest polydispersity of the polymers. In the polymer melts, a dependence of D on M of the form of the Eq. (7) was observed, with an exponent of —1.7. Figure 3 shows the principal data.

Cosgrove and Warren [33] used the PGSE method of Packer, et al. [9]. with the Meiboom-Gill rf pulse sequence [7] to investigate concentrated polystyrene solutions.

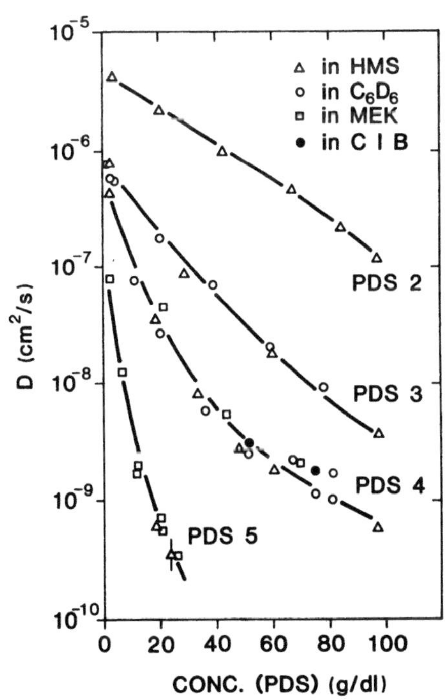

Fig. 3. Self-diffusion of polydimethylsiloxanes in various solvents as function of concentration. Polymer molecular weights varied from 4.6×10^3 (PDS2, top) to 8×10^5 (PDS5, bottom). (after Ref. [40], with permission).

They compared Eq. (7) to their data for molecular weights M less than M_c, the value where entanglements set in, and obtained n = —0.53 for a 30% polymer solution and n = —0.7 for a 40% solution. A much smaller value of n was obtained above M_c — contrary to predictions of the theory of reptation[26] — and attributed to contributions by segmental motions to the measured diffusion coefficients. However, their analogy of the PGSE and neutron scattering experiments appears flawed by an apparent misunderstanding of the difference in time scales provided by the two experiments.

A PGSE study of nearly monodisperse cis-polyisoprene melts below M_c ($400 \leqq M \leqq 10^4$) at various temperatures (23 °C ≦ T ≦ 100 °C) was reported by von Meerwall, et al.[41]. They found that Eq. (7) would not fit their data, firstly because of an upward concavity in the plot of log D *vs.* log M, and secondly because both slope and curvature were more pronounced at lower temperatures. Instead, they showed that a Rouse-like scaling behavior (Eq. (7) with n = —1) could be made to fit the data provided that the reduction of the fractional free volume f with increasing M was taken into account. Thus, following Bueche[42],

$$D(T, M) = PM^{-1} \exp(-B_d/f), \tag{8}$$

with P independent of M, T, and f. The latter takes the form[42]

$$f = f_g + \Delta\alpha(T - T_{g\infty}) + V_E\varrho M^{-1}; \qquad (T \geqq T_g). \tag{9}$$

Here B_d is a dimensionless constant of order unity[43], ϱ being the polymer density; f_g (≈ 0.025) is the fractional free volume at the glass transition temperature T_g, with the latter taking its maximum value $T_{g\infty}$ at infinite molecular weight. V_E denotes the

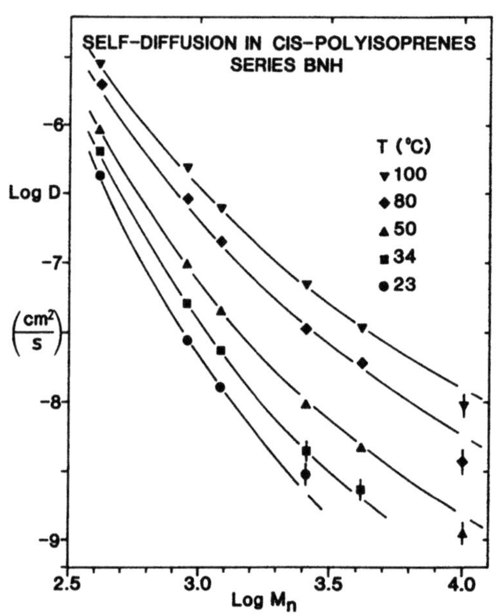

Fig. 4. Self-diffusion in cis-polyisoprene melts as function of molecular weight, at five temperatures. Curves are single fit of Eq. (8) with Eq. (9) to all data. (Ref.[41], with permission).

extra free volume associated with the ends of a polymer molecule, these being more abundant at low M. $\Delta\alpha$ is the thermal expansivity of the remaining free volume.

It was possible to obtain an excellent fit of Eq. (8) with Eq. (9) to the combined M and T dependence of D and extract of value $V_E = 13.7 \pm 0.2\ cm^3/mol$, slightly more in a similar polymer having one terminal —OH group. The constants P and f_g had to be fitted as well, the latter assuming the value $f_g = 0.023 \pm 0.001$. This interpretation of melt diffusivity was confirmed in separate experiments in which diffusion of the smaller oligomers was measured in a high molecular weight cispolyisoprene host (see below). Data and fitted theory in the melts are shown in Fig. 4.

Star-branched cis-polyisoprene and polystyrene molecules in solution with CCl_4 and C_6F_5Cl were examined by von Meerwall et al. [32] via PGSE and NMR relaxation. In a comparison of the diffusion of linear and equi-armed star-branched molecules over the entire concentration range, it was found that for unentangled or nearly unentangled ($M_{star} < 10^5$) molecules the effect of star-branching was virtually undetectable provided total molecular weight was kept constant. (An exception was the behavior of the 8 and 18-armed stars in the dilute regime, a subject which will be discussed below.) Thus in concentrated solutions and melts in the absence of entanglements, molecular architecture is relatively unimportant to diffusion, which is dominated by the monomeric friction coefficient and the degree of polymerization [44]. There was

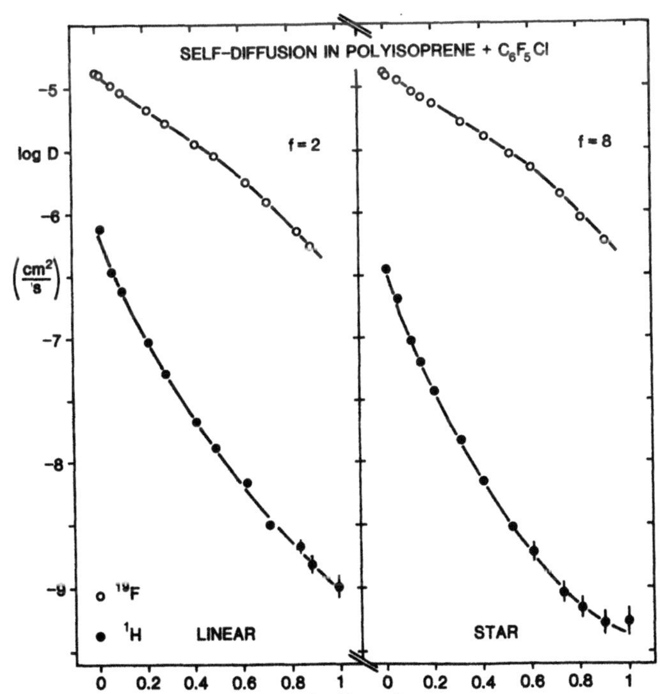

Fig. 5. Diffusion of C_6F_5Cl (open symbols) and cispolyisoprene (filled symbols) in solution at 50 °C, from dilute solution to the melt. Left: linear polymer, $M = 10^4$; right: eight-armed star-branched polymer, $M = 4 \times 10^4$ (Ref. [32], with permission).

indirect evidence that entanglements are less frequent among star-branched molecules than among linear molecules of the same molecular weight. At the highest molecular weights and polymer concentrations, diffusion coefficients well below $D = 10^{-9}$ cm^2 per sec were observed. Simple calculation *via* Eq. (5) showed that in these cases the value of D measured by PGSE methods probably no longer reflects pure center-of-mass motion of the molecules, but may include contributions from segmental mechanisms. Figure 5 shows representative data from this work.

5 Polymers in Dilute and Semidilute Solutions

In their investigation of polydimethylsiloxane and poly(ethylene oxide) in solution with various solvents, Tanner, Liu, and Anderson[40] extrapolated the observed polymer diffusion coefficients to zero polymer concentration c. They applied Flory's theory of dilute solutions[45] to the case of diffusion:

$$D_0 \equiv D(c = 0) = \frac{kT}{\eta_0 P M^{0.5}} \frac{1}{\alpha} \left[\frac{M}{\langle s^2 \rangle}\right]^{0.5}, \tag{10}$$

where P is a numerical constant, η_0 is the solvent viscosity, and α represents the solvent power, a quantity approximately proportional to $M^{0.1}$ for good solvents but independent of M for theta solvents. The term in square brackets is the ratio of the molecular weight to the mean square dimension of the coiled polymer molecule at infinite dilution in the solvent, and, for a given polymer chemical structure and molecular architecture, becomes independent of M above some minimum value on the order of $M \approx 10^3$. Eq. (10) thus implies that the molecular-weight dependence of D_0 should follow a scaling-law of the form of Eq. (7), with n between -0.5 and -0.6 depending on the solvent power. In fact, Tanner, Liu, and Anderson obtained $n = -0.6$ in most of their solvents over a wide range of molecular weight. Comparison with intrinsic viscosities in the same systems produced satisfactory agreement except at high molecular weights.

Moseley[46] used a highly automated PGSE NMR spectrometer to measure diffusion of short-chain polystyrenes in deuterochloroform solutions. Since the molecular weights were relatively small ($510 \leqq M \leqq 3300$), the solution behavior was characteristic of dilute and semidilute solutions even at polymer weight fractions approaching 0.5. Thus, plots of log D vs. c displayed linear decreases with increasing c, the slope becoming steeper with increasing M. D_0 appeared to depend on M in the manner suggested by Eq. (10) for good solvents.

Pimenov et al.[47] reported PGSE diffusion measurements of linear polystyrenes ($2.5 \times 10^4 \leqq M \leqq 3.6 \times 10^5$) in dilute and semidilute solution in CCl_4. They, too, found a decrease of D with increasing polymer concentration and increasing molecular weight. In an effort to detect a power-law in the molecular weight dependence of D they plotted log D vs. log M at a fixed polymer concentration of 100 kg/m^3, corresponding to a semidilute solution at the lowest M. The observed behavior, however, failed to conform to a straight line, displaying substantial downward concavity, the reason for which may be that at higher M the fixed polymer concentration apparently lies in the concentrated regime with its more pronounced concentration dependence.

Callaghan and Pinder [48,49] used the PGSE method in a detailed examination of the diffusion of linear polystyrene molecules dissolved in CCl_4. They applied standard dilute hydrodynamic theory to self-diffusion (as distinct from mutual diffusion) and identified the lowest-order concentration dependence of D with the coefficient k_F, writing

$$D^{-1} = D_0^{-1}(1 + k_F c + \ldots) . \tag{11}$$

They obtained a scaling behavior $D_0(M)$ characteristic of inferior solvents (see Eq. 10): $n = -0.51 \pm 0.02$. Plots of D^{-1} vs. c were linear up to more than 60 kg of polymer per m^3 of solution at low M ($M \leqq 3 \times 10^4$) (see Fig. 6). The dependence of k_F on M was explained by the hydrodynamic model of Pyun and Fixman [50] at lower M, and by that of Yamakawa [51] at the highest molecular weights. The Pyun-Fixman model, as elaborated by King et al. [52], permits relating k_F to D_0 and thus to M:

$$k_F \propto D_0^{-3} M^{-1} \propto M^{-1-3n} \tag{12}$$

given that $D_0 \propto M^n$ (i.e., e.q. 10). The Yamakawa model contains a term of higher power in M and thus does not result in a pure power law.

The authors semiquantitatively described a semidilute concentration regime with limits

$$c^*(M) < c < c^{**}(M) . \tag{13}$$

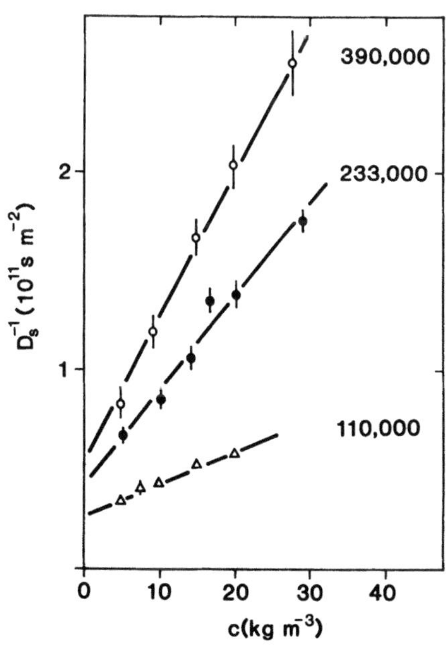

Fig. 6. Reciprocal diffusivity versus concentration in dilute solutions of linear polystyrenes in CCl_4, for three polymer molecular weights (indicated) (after Ref. [49], with permission).

At the lowest molecular weights ($M \lesssim 2 \times 10^3$) the first inequality could not be satisfied at any concentration, but at higher M this regime should have finite extent. Rather than identifying this region with unentangled but hydrodynamically screened semidilute behavior [44], Callaghan and Pinder [48] interpreted it in terms of gel-like behavior for which deGennes had predicted [26]

$$D(c) = D'(M)\, c^{-1.75}\,. \tag{14}$$

In fact, their data for $M \geqq 10^5$ displayed scaling behavior of this sort, although even in their smallest polystyrenes a tangent of slope —1.75 could be drawn to plots of log D vs. log c.

Further, reptation theory asserts [26] that the molecular-weight dependence of the diffusion coefficient in an entangled gel should have the form

$$D(c') \propto M^{-2}\,. \tag{15}$$

Callaghan and Pinder, taking c′ to be a fixed multiple of their estimated c*, obtained such scaling behavior, but with an exponent of —1.4 rather than —2.

Lastly, Callaghan and Pinder believed to have detected cooperative diffusion in their experiment [48]. Their polystyrene specimen of $M = 2 \times 10^6$ displayed a two-component echo attenuation, interpreted in terms of two time-dependent diffusion coefficients. The more rapid component was not observable at long diffusion times, and at short diffusion time (t = 16 ms) scaled with polymer concentration (2, 4, and 8%) in the manner

$$D_c(c) \propto c^{+0.75}\,, \tag{16}$$

consistent with deGennes' prediction [26] for gel-like cooperative segmental displacements.

A comparison of dilute and semidilute diffusion behavior between linear (F = 2) and three-armed star (F = 3) polybutadienes and polystyrenes in CCl_4 was recently reported by von Meerwall et al. [53]. They found that in their range $2 \times 10^3 < M < 10^6$ there were no significant differences between F = 2 and F = 3 for a given polymer at the same total molecular weight, in accordance with an observation in earlier similar work [32] but at variance with the results of intrinsic vicsocity measurements in these systems. This identity included D_0, k_F, as well as semidilute diffusivities. Significant differences between polystyrenes and polybutadienes (each including both architectures) existed in all three of these measures. In particular, D_0 scaled with M(star) to the power -0.52 ± 0.01 for polystyrenes — in agreement with other work [49] — but to the power -0.60 ± 0.01 for polybutadienes. The Pyun-Fixman model (Eq. 12) was a reliable guide to the molecular-weight dependence of the measured k_F for both polymers. Instrinsic viscosity measurements confirmed the conclusion that CCl_4 is a substantially better solvent for polybutadienes than for polystyrenes. Data for D_0 vs. M is shown in Fig. 7.

As part of this work [53], the authors reproduced and confirmed many of the observations of Callaghan and Pinder [48,49] on linear polystyrenes in CCl_4. A few differences were evident, however, beyond the quantitative offsets arising from

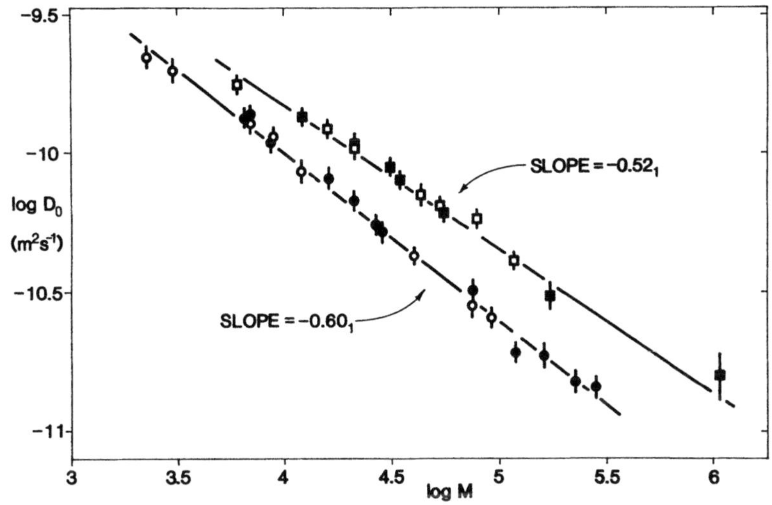

Fig. 7. Self-diffusion of linear (open symbols) and three-armed star (filled symbols) polystyrenes (squares) and polybutadienes (circles) in CCl_4 extrapolated to infinite dilution, as function of polymer molecular weight (Ref. [53]), with permission).

a temperature difference of some 22 °C between these experiments. Strong upward curvatures in plots of 1/D vs. c were observed above $c \approx 15$ kg/m^3 at molecular weights above 10^5. The effects of contamination by trace solvents was clearly evident and had to be carefully eliminated at the source. After this was done, essentially single-component time-independent diffusion was observed in all samples; a small distribution of diffusion coefficients was successfully modelled in terms of the known polydispersity ($\bar{M}_w/\bar{M}_n \approx 1.06$) of the polymers. In particular, a three-armed polystyrene star of $M = 10^6$ was no exception to this orthodox behavior, whereas a linear polystyrene preparation of $M = 2 \times 10^6$ examined by Callaghan and Pinder [48]) had exhibited two-component time-dependent diffusion, which had been interpreted in terms of gel-like cooperative motions. Thus at this writing it is not clear whether the adherence of Callaghan and Pinder's data for one specimen to Eq. (16) is incidental to the intended significance of this theory.

In their PGSE investigation of star-branched polymers (cis-polyisoprene and polystyrene) in CCl_4 and C_6F_5Cl, von Meerwall et al. [32]) examined diffusion in the dilute limit as function of F, the number of molecular arms of equal length. For cis-polyisoprene, F = 2(linear), 3, 8, and 18 were studied mainly under conditions of constant arm molecular weight $M(\text{arm}) = 5 \times 10^3$. In semidilute and concentrated solutions as well as the melt, no distinction could be found in the diffusive motion on the basis of F beyond differences attributable simply to changes in M(star). Since reptation in entangled gels and melts should be severely inhibited by star-branching, it was concluded that the solutions and melts were not sufficiently entangled to display reptation and tube renewal or reorganization, even though the molecular weight for some star molecules approached $M = 10^5$. The existence of two-component spin-echo attenuation at higher concentrations and in the melts was shown to arise from minute traces of solvents and antioxidants used in the preparation and characterization of the polymers. The fast-decaying echo component could be reduced or eliminated by heating and evacuation of the specimen before redissolving. By pursuing echo attenuation to about 1/10 or 1/20 of the original height, a modest range ($D_{max}/D_{min} \lesssim 2$) of diffusivity was detected at all concentrations and shown to orginate in the small polydispersity of the polymers ($\bar{M}_w/\bar{M}_n \lesssim 1.08$).

In the limit of infinite dilution, one distinction between the F-dependence and M-dependence of D_0 did become evident. Since M(star) is proportional to F, i.e.

$$M(star) = F \cdot M(arm)\,,$$

a plot of log D_0 vs. log F at constant M(arm) should yield a slope of —0.5 to —0.6 (see Eq. 10). However, the observed slope of this plot is rather smaller (negatively) than —0.5, a fact attributable in part to the change in architecture. Also, the ease in draining of the solvent is impaired by the increase in segment density at high F; small molecules with large F moreover tend to be relatively more expanded due to crowding of the inner segments, and to chain stiffness effects. The concentration-dependence of D in the dilute regime (e.g., k_F) no longer depended on F and M(arm) separately, but only on their product M(star) for any F.

The calculation of friction coefficients from self-diffusion experiments may give results substantially at variance with friction coefficients deduced from sedimentation or mutual diffusion experiments; agreement is only obtained in the limit of zero polymer concentration. This point is made in two investigations using PGSE methods to measure self-diffusion, one [54] of polystyrene ($M = 1.1 \times 10^5$) in deutero-toluene at 25 °C, the other [55] of dextran ($M = 4.4 \times 10^4$, 6.4×10^4) in deuterated water at 75 °C. Application of theoretical scaling arguments to both sedimentation and self-diffusion confirms the findings that the sedimentational friction coefficient increases with increasing polymer concentration considerably more rapidly than the diffusional friction coefficient, in the dilute as well as the semidilute regimes.

6 Light Penetrants and Diluents in Polymers

While to some extent the boundary between light and heavy penetrant molecules in polymers appears arbitrary, theoretical explanations of the transport behavior of the latter tend to be more complex.

Numerous NMR studies [1,2] have reported the effect of diluent molecules on the molecular motion of the host polymer, and a number of investigations have reported on transport and migration of molecules dissolved in polymers near and above their glass transition temperature T_g. The earliest measurements of diluent self-diffusion via SGSE methods appear to be those of McCall, Douglass, and Anderson [17], who in their study of polyisobutylene transport comment that the diffusional activation energies of polymer and solvent tended to be similar in magnitude, a point later elaborated theoretically for liquid solutions [56].

In a search for evidence of restricted diffusion, Woessner [57] used standard SGSE techniques to measure diffusion of (*inter alia*) benzene in a filled and crosslinked rubber; he interpreted his data as demonstrating the existence of restrictions. In his dissertation, Tanner [6] showed that the PGSE experiment, because of its more precise experimental definition or diffusion time, is much better suited for studying restricted (or any other non-Fickian) diffusion. He repeated Woessner's experiment but failed to find evidence of restricted diffusion on the distance scale (i.e., a few μm) to which spin-echo experiments are sensitive.

In his PGSE study of polyethylene oxide and polydimethylsiloxane in C_6H_6 and $CHCl_3$, Tanner [39] also measured the diffusion of a fixed fraction of solvent in the polymers. He concluded that their diffusion rate in polymers of molecular weight lower than their own was approximately equal to that of the polymers. As the polymer molecular weight exceeded that of the solvent, the solvent diffusion rate approached a constant value, independent of polymer molecular weight. Tanner offered semiempirical explanations for this effect.

Another instance in which solvent diffusion rate is independent of the (much greater) host molecular weight was observed in two polyisoprene — C_6F_5Cl systems across the entire concentration range [32]. The two molecular species differed in molecular weight by a factor of four, and in architecture in that one was linear, the other an eight-armed star-branched molecule. Neither the difference in host molecular weight nor the infrequent branching in the star molecules had any effect on solvent diffusion, which was identical at equal concentrations; polymer diffusion at equal concentrations differed significantly but was at least one order of magnitude slower than solvent diffusion.

Boss, Stejskal and Ferry [29] used the PGSE method to investigate polyisobutylene-benzene solutions. Employing the proton resonance they observed a spin echo consisting of contributions from both polymer and solvent. Because the

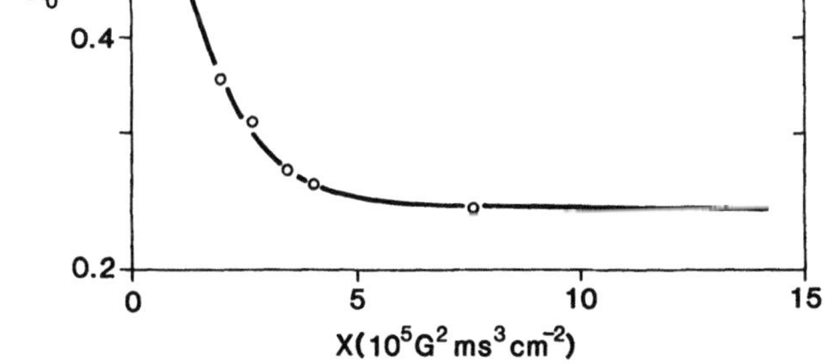

Fig. 8. Spin-echo amplitude A(δG)/A(0) vs. x (see Eq. 2) in a poly-isobutylene-benzene solution containing 59 vol. % benzene. Initial decay rate is due to solvent diffusion, slower decay describes polymer diffusion; figure shows only the lowest 1/4 of the abscissa domain used in fitting the double exponential (after Ref. [29], with permission).

diffusivities of the two species differed greatly, they were able to resolve these by fitting a double exponential to the echo attenuation curve (see Fig. 8); however, they preferred to use a three-pulse sequence to eliminate the undesired polymer echo when measuring diluent diffusion.

These authors were the first FGSE workers to make extensive use of the concept of free volume [42,44] and its effect on transport in polymer systems. That theory asserts that amorphous materials (liquids, polymers) above their glass transition temperature T_g contain unoccupied volume randomly distributed and in parcels of sufficient size to permit jumps of small molecules — and of polymer jumping segments — to take place. Since liquids have a fractional free volume f_{dil} typically greater than that, f_p, of polymers, the diffusion rate both of diluent molecules and (uncrosslinked and unentangled) polymer molecules should increase with increasing diluent volume fraction v_{dil}. The Fujita-Doolittle expression [43] describes this effect quantitatively for the diluent diffusion:

$$\log D_1(v_{dil}) = \log D_1^0 + \frac{B_d s v_{dil}}{2.3(1 + s f_p v_{dil})}, \tag{17}$$

$$\text{where } s = \frac{f_{dil} - f_p}{f_p^2}, \quad B_d \approx 1$$

Boss, et al., fitted Eq. (17) to their data D_1 vs. v_{dil}, enabling them to determine f_p and D_1^0. At solvent concentration approaching $v_{dil} = 0.95$, the data revealed an enhancement above the value predicted by Eq. (17) as fitted to the lower-concentration data. The authors argued that under these circumstances macroscopic inhomogeneities in concentration (and hence in the free-volume distribution) should exist and enhance the diffusivity. Above $v > 0.99$ the polymer coils no longer overlapped substantially, depriving the solvent molecules of a set of obstacles fixed with respect to the laboratory, and solvent diffusion became related principally to intrinsic viscosity.

Kosfeld and Goffloo [58] used an alternating-gradient modification of the PGSE method to investigate solvent diffusion in the systems benzene-polystyrene and cyclohexane-polystyrene at polymer concentrations below 20–30 wt. %. They examined temperature as well as concentration dependences, utilizing free-volume theories to interpret their data. As is generally observed, diffusional activation energies $E_a = -R\,\partial \ln D/\partial T^{-1}$ increased with increasing polymer content. The concentration dependences log D vs. polymer weight fraction showed the characteristic downward concavity, but any diffusivity enhancement near pure solvent seemed to be small or obscured. Application of the Fujita-Doolittle equation (Eq. 17) to the data yielded D_1^0 (for the solvent) and f_p (for polystyrene). These parameters in both systems showed orthodox behavior except near (50 ± 10°) C, where they were notably enhanced; this behavior was attributed to a rapid increase in polystyrene intramolecular mobility, i.e. displacements of the phenyl groups. The 50% difference in f_p as deduced from experiments with different solvents was explained via the concept of a van der Waals volume differing between solvents, although differences in f_{dil} and B_d may have played a part.

The temperature-dependence of solvent diffusion obeyed the free-volume — based expression of Williams, Landel, and Ferry [59] adapted to diffusion:

$$\log D = \log D_g + \frac{C_1(T - T_g)}{2.3(C_2 + T - T_g)} \tag{18}$$

with the identification

$$C_1 = B_d/f_g;\ C_2 = f_g/\Delta\alpha$$

in the notation defined earlier (see Eqs. 8, 9). Application of Eqs. (17) and (18) to the extrapolations of D to infinite solvent dilution permitted Kosfeld and Goffloo to extract values for $f_p(T = T_g)$ and $\Delta\alpha$(PS). Eq. (18) was, in fact, obeyed for both systems above 70 °C and below 40 °C.

Polyvinylchloride was the host polymer in a study of the diffusion of dimethylphthalate, dibutylphthalate, and dioctylphthalate, performed by Maklakov, Smechko, and Maklakov [60] between room temperature and 110 °C. Azancheev and Maklakov [61] extended this work to include polystyrene as host, and to dependences of diffusion on concentration. They concluded that the macromolecules did constrain and "trap" the phthalate molecules at high polymer concentration, but without inhibiting the mobility of these diluents at lower polymer concentrations, e.g., in the gel. They used a version of the free volume theory to give a semi-quantitative explanation of the temperature and molecular size dependence of phthalate diffusion.

Zupančič, et al. [62] performed an NMR and PGSE study of the motion of butane in linear polyethylene as function of gas pressure. They considered the polymer, a semi-crystalline powder, to be composed of stacks of parallel lamellae in an amorphous matrix. This topology resulted in a "detour factor" of 1/3; no absolute blocking occurs. The authors demonstrated that in the presence of topological constraints PGSE experiments offer a particular simplicity of interpretation by comparison with most bulk techniques for measuring diffusion. In this case, the free diffusion rate in the medium between lamellae was simply three times the (single) measured rate. Butane concentration in the polymer being proportional to pressure, log D increased linearly with $v_{dil}(v_{dil} \ll 1)$; such behavior is expected on the basis of Eq. (17) in the limit of low v_{dil} (see Fig. 9).

The first demonstration of PGSE measurements using ^{13}C NMR was given by Moseley and Stilbs [63]. Because of the low signal-to-noise ratio and long T_1 relaxation time, Fourier transform techniques combined with signal averaging were essential. Diffusion of trans-decalin in polystyrene was measured as function of concentration at several temperatures. Nyström et al. [64] reported a phenomenological study of the diffusion of dichloromethane and cyclopentane in polystyrene as function of concentration, noting significant differences between these solvents.

An industrial oil-extended synthetic cis-polyisoprene was investigated by von Meerwall and Ferguson [30]. Following Boss, et al. [29], they substracted the unattenuatable spin echo arising from the rubber, obtaining the diffusivity of the extender oil from the remainder. They demonstrated that no departures from Fickian diffusion occur, and measured the diffusion of the oil, both in the rubber and in the pure liquid, between −10 °C and 130 °C. Since the plot of log D vs. 1/T was not a straight line it was necessary to invoke the Williams-Landel-Ferry temperature dependence,

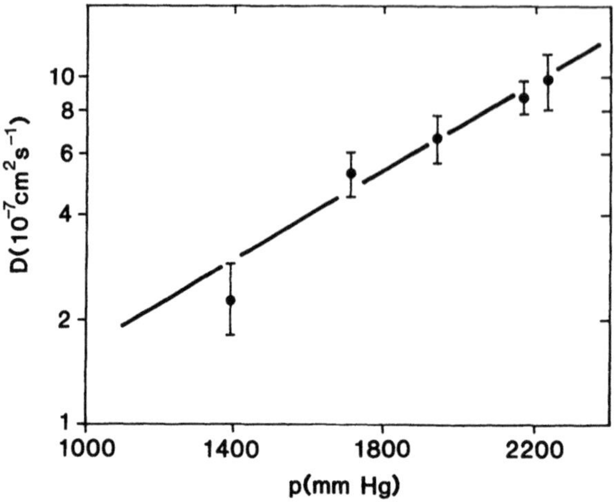

Fig. 9. Pressure-dependence of butane diffusion in linear polyethylene at room temperature. (after Ref. [62], with permission).

Eq. (18). Their experiment provided an opportunity only to fit three of the four unknown constants, namely D_g, c_2, and T_g. T_g in the solution was found to be depressed below that known for the rubber. An examination of the concentration dependence of oil diffusion in the same rubber host confirmed the applicability of Eq. (17) below $v_{dil} \lesssim 0.9$ and permitted a measurement of the fractional free volume of one or both components of the oil-rubber solution.

The first PGSE investigation of a rubber-based ternary solution was described by Ferguson and von Meerwall [31], who measured diffusion of C_6F_6(^{19}F NMR) and n-paraffin (n-dodecane or n-hexatriacontane; ^1H NMR) in a commercial polybutadiene as function of both concentrations. They showed that both concentration dependences in the ternary region can be derived from the measured diffusivity of each diluent i = 1, 2 in binary solution in the rubber. To do this it was necessary to extend the Fujita-Doolittle expression, as follows:

$$
\begin{aligned}
v_{dil} &= v_1 + v_2 , \\
f_{dil} &= (v_1 f_1 + v_2 f_2)/v_{dil} , \\
\log D_i &= \log D_i^0 + \frac{B_{di} s v_{dil}}{2.3(1 + f_p s v_{dil})} ,
\end{aligned}
\qquad (19)
$$

with s defined as in Eq. (17). The manner of extension of Eq. (17), by its success in reproducing the ternary data, implied that for free-volume purposes both diluents form a common liquid sharing a volume-averaged f_{dil}. Other aspects of this work will be discussed below.

In their study of the NMR T_1 and T_2 of crosslinked cis-polyisoprene sheets under extension, von Meerwall and Ferguson [65] found that T_2 of the rubber had much smaller anisotropy ("magic angle" effect) than that of trace penetrants at the same extension ratio $\lambda < 3$. However, the penetrant diffusion (referred to the strained dimensions) was within experimental error isotropic; these findings are equally valid for C_6F_6 and n-hexadecane as penetrant. The authors concluded that segment orientational order sufficient to give rise to measurable anisotropy of penetrant diffusion is probably not attainable at realizable extension ratios.

Diffusion of 1,3-dimethyladamantane, a rigid molecule of intermediate size, was measured via PGSE methods in a polybutadiene at various temperatures over the complete concentration range [66]. Diffusion of both components in the liquid solution of 1.3 DMA + C_6F_6 was also measured. Von Meerwall and Van Antwerp found that the Williams-Landel-Ferry expression (Eq. 18) for temperature dependence held at all concentrations in the rubber-based solution, but that the Fujita-Doolittle theory (Eq. 17) was unable to reproduce the concentration dependences. They found it necessary to invoke the free-volume expression derived by Vrentas and Duda [67] to generalize the Fujita-Doolittle theory. It is expressed in terms of weight fraction w_i of component i; in a two-component system (i = 1, 2), the theory may be written

$$\log D_i = \log D_i^0 + \frac{B_i w_1 s_i}{2.3(1 + w_i f_2 s)}, \; i = 1, 2$$

$$\begin{aligned} \text{where } s &= (\varrho_2 f_1/\varrho_1 - f_2)/f_2^2; \\ s_1 &= (\varrho_2 m_1 f_1/\varrho_1 m_2 - f_2)/f_2^2, \\ s_2 &= s_1 m_2/m_1, \text{ and} \\ B_i &= g\hat{V}_i^* \varrho_2 . \end{aligned} \qquad (20)$$

Here ϱ denotes density, m represents collision-dynamic mass (which may depart substantially from the molecular mass), $\hat{V}_i^*$ is the specific critical volume for diffusion of species i, and g is an overlap factor $0.5 \leq g \leq 1$. The fractional free volumes f_i

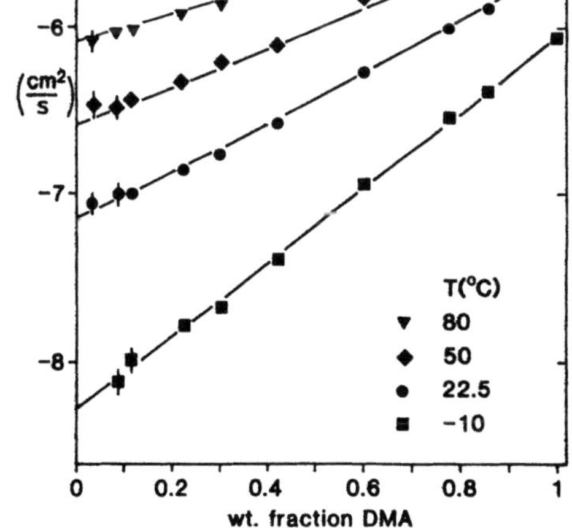

Fig. 10. Diffusion of 1,3 dimethyladamantane in polybutadiene as function of concentration, at four temperatures. Lines are single fit of Eq. (20) with Eq. (9) to data (Ref. [66], with permission).

were assumed to have the usual temperature dependence of Eq. (9). The number of unknown parameters was sufficiently high that information from three experiments (C_6F_6 — rubber: w_1 — dependence; 1,3 DMA-rubber: w_1 and T dependence; 1,3-DMA-C_6F_6: w_1 dependence of D_1 and D_2) had to be combined. The results for the DMA-rubber system suggested that this diluent molecule (M.W. = 164) is small enough not to be effectively bulky since the host rubber molecular segments can still accommodate themselves to its passage: both B_1 and m_1/m_2 are near unity. In solution with C_6F_6 the observed non-ideality ($B_1m_1/B_2m_2 \neq 1$) was shown to originate in differences in molecular geometry. Figure 10 shows the DMA-rubber work.

Polymer gels and polyelectrolyte solutions are attractive subjects of study with the PGSE method. In their pulsed-NMR investigation of agarose gels, Derbyshire and Duff[68] also studied the diffusion of the free water. Their PGSE data for gel concentrations of up to 18% was characterized by a single diffusion rate with no evidence of restricted diffusion. The decrease of D with increasing gel concentration was approximately linear and consistent with a model of closely spaced topological constraints by the hydrated macromolecules.

Strong evidence of ionic association was found by Stilbs and Lindman[69] in their PGSE study of aqueous polyelectrolyte solutions, polyacrylic acid and polymethacrylic acid, neutralized by tetramethylammonium hydroxide, with or without sodium counterions. While polymer diffusion could not be detected since its T_2 was too short, TMAOH and water diffusion was measured as function of degree of neutralization α', or Na^+ content. A pronounced minimum of D(TMAOH) near $\alpha' = 1$ was interpreted in terms of a two-site model, leading to the determination that at $\alpha' = 1$, approximately half of the counterions are bound in both systems. Fourier transform techniques permitted the simultaneous measurement of diffusion of water and TMAOH.

7 Large or Flexible Molecules Dissolved in Polymers

A systematic study of n-paraffins (n-octane through n-hexatriacontane) diffusing in various rubbers was reported by von Meerwall and Ferguson[70], who principally measured the concentration dependences below 60 wt. % paraffins. They showed that the Fujita-Doolittle equation applied and, given f_p from literature data, extracted f_{dil} and D_0 from their measurements. Converting f_{dil} to molar free volume V_f they found that V_f initially rises linearly with carbon number N, yielding a value of $2V_E \approx 15\ cm^3\ mol^{-1}$ and $dV_f/dN = 1.5\ cm^3\ mol^{-1}$ per monomer unit. At $N \gtrsim 24$ the increase ceased, a fact which was interpreted in terms of the onset of segmental mobility. By comparing data in several host rubbers the authors confirmed that B_d (see Eq. 17) for paraffins in cis-polyisoprene is anomalously low, near 0.67. Of equal interest were the intercepts D_0, which were shown to scale inversely with paraffin molecular weight in two host rubbers for the entire series of paraffins. The latter observation reemphasizes the dependences of transport on the (nearly constant) monomeric friction coefficient and the degree of "polymerization" even during the transition to segmental mobility in the absence of entanglements with the host species (which itself was strongly entangled). Figure 11 shows the results in polyisoprene solutions.

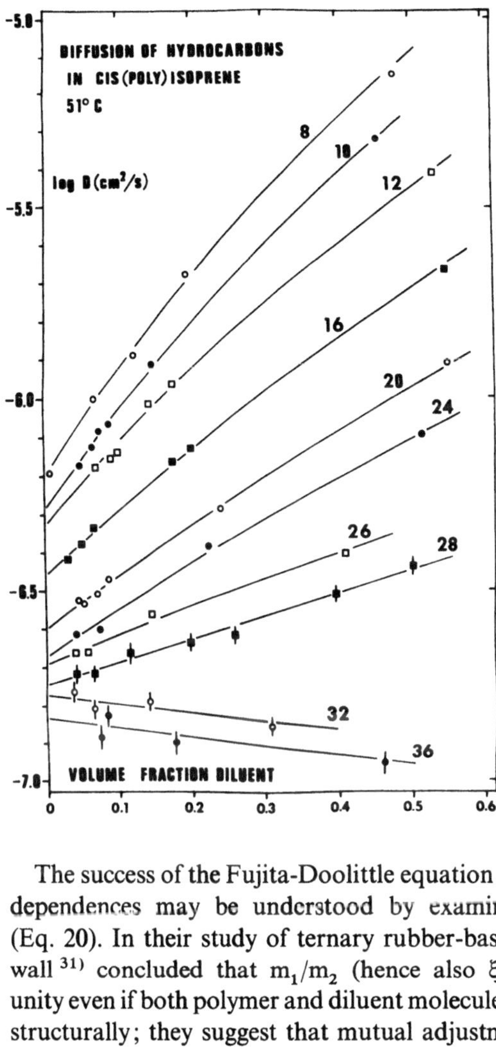

Fig. 11. Diffusion of n-paraffins in a cis-polyisoprene as function of concentration at 51 °C. Lines are two-parameter fits of Eq. (17) to data. Paraffin carbon numbers are indicated. (Ref. [70], with permission).

The success of the Fujita-Doolittle equation in explaining diffusional concentration dependences may be understood by examining its Vrentas-Duda generalization (Eq. 20). In their study of ternary rubber-based solutions Ferguson and von Meerwall [31] concluded that m_1/m_2 (hence also $\xi \equiv B_1m_1/B_2m_2$) tend to remain near unity even if both polymer and diluent molecules are segmentally mobile and dissimilar structurally; they suggest that mutual adjustments between colliding segments keep their collision-dynamic effective masses approximately equal. There was, however, evidence that this near-equality is no longer maintained if the diluent molecule is substantially lighter or smaller than a polymer jumping segment; the Fujita-Doolittle equation, which implicitly assumes $m_1/m_2 = 1$, finally breaks down in these cases.

A similar breakdown occurs in non-polymeric liquid solutions of two greatly unlike molecules. In their study of the three binary solutions among 5-alpha-cholestane (M.W. 373), C_6F_6, and polybutadiene, von Meerwall and Ferguson [71] found one striking unorthodoxy in the C_6F_6—5αC solution. There the deduced ratio m_1/m_2 differed by a factor of four from the molecular weight ratio, showing that several C_6F_6 molecules are involved in a cooperative displacement to permit the passage of a single cholestane molecule. (In the system C_6F_6—1,3 dimethyladamantane, m_1/m_2 had still approximated the molecular weight ratio [66]). Also, the diffusional activation energy of both components doubles between 60 and 100 wt. % cholestane in C_6F_6. However, even the rather bulky cholestane molecule still found relatively ready accommodation in polybutadiene: at 80 °C, m_1/m_2 was no larger (~1.4) than it was in the dimethyladamantane-polybutadiene system [66]. The only diffusional attribute of the cholestane-polybutadiene system reminiscent of molecular bulk relates not directly to the size but to the low fractional free volume of cholestane. Its diffusivity in the rubber decreased with increasing concentration, by nearly a factor of two between trace concentration and the pure liquid, thus displaying pronounced antiplasticizing behavior. Similar but smaller effects had been observed in the diffusion of heavier n-paraffins in two rubbers [70]. By comparison, B_1 (or else B_1/B_2; see Eq. 20) appeared to display no peculiarity related to molecular size, either in liquid or polymer-based solutions, being mainly related to molecular geometry and its influence on the efficiency with which local free volume can be utilized for diffusional jumps.

Segmental mobility appeared to have no direct effect on m_1/m_2 or on B_1, whether the macromolecules dissolved in entangled melts were foreign [70] or oligomers of the host molecule [41]. (The indirect effect on the free volume was mentioned above.) Similar findings apply to smaller macromolecules dissolved in liquids in non-dilute solutions [37]. In fact, diffusion of oligomers in their high polymer cis-polyisoprene hosts was sufficiently simple in its concentration dependence [41] to permit the use of Eq. (17), from whose slope s the chain-end contribution to free volume was determined. The molecular-weight-dependence of diffusion of trace macromolecules tends to follow Eq. (7) with an exponent $n = -1$ if no entanglements with host molecules occur [70]. However, comparison of the diffusion rate of two monodisperse cispoly-isoprene oligomers (M = 411, 895) in an entangled but uncrosslinked cispolyisoprene host [41] yielded an exponent $n = -1.5 \pm 0.1$. At present it is not clear whether this departure from $n = -1$ reflects a transition into the reptation regime ($n = -2$ or greater), the effect of localization of chain-end free volume due to inefficient packing at low M [44], or some alternate cause.

8 Conclusions and Outlook

The free volume theory has been eminently successful in correlating self-diffusion in polymers, particularly the temperature and concentration dependences in polymer-solvent systems. However, intermolecular attraction is known [72] to play a significant role in inhibiting diffusion; naive application of Eq. (17) or Eq. (20) across the full concentration range may distort the deduced free volume parameters. But neither the free volume theory nor the scaling approach provide a basis for predicting absolute diffusion rates under given conditions. Theories which do attempt to predict absolute rates based on molecular and dynamic parameters appear not yet to be sufficiently mature to be employed in quantitative assessments of rate information to the needed precision. Thus, parameters such as K(T) in Eq. (7), or D_0 in Eq. (17) and Eq. (20), etc., are generally merely empirically correlated with molecular and structural attributes [73].

The field gradient spin echo methods appear now to have amply proven their validity and usefulness in the study of polymer systems. This early work has been understandably academic in flavor, but there seems to be no impediment to the use of FGSE measurements for routine rheological characterization, e.g., to supplement viscosity measurements. Particularly for industrial corporations which already perform NMR measurements, a relatively minor addition to an existing spectrometer can greatly increase its usefulness by permitting rapid selfdiffusion measurements.

Of particular interest in practical polymer work are cases where several substances diffuse simultaneously, or where diffusion is anisotropic or inhomogeneous, as in partially crystalline or filled rubbery polymers. For such cases PGSE measurements offer their greatest advantages, hitherto apparently not fully exploited, although similar problems in biological systems are being vigorously attacked with this method. Thus polymer solutions, gels, and emulsions are excellent candidates for study with the PGSE method. The only severe limitation of the method is the relatively modest lower limit for the measurable diffusivity; no more than another order of magnitude (to $D \gtrsim 10^{-11}\ cm^2\ s^{-1}$) can be reasonably expected to be gained in optimal cases through the use of pulse sequences which elicit spin echoes at long diffusion times.

9 References

1. For recent reviews of the pertinent literature, see V. J. McBrierty and D. C. Douglass: Phys. Reports (Phys. Letters) *63*, 61 (1980); also J. Polym. Sci. Macromol. Revs. *16*, 295 (1981)
2. Slonim, I. Ya., Liubimov, A. N.: The NMR of Polymers (transl.) Plenum Press, New York (1970)
3. Hahn, E. L.: Phys. Rev. *80*, 580 (1950)
4. Carr, H. Y., Purcell, E. M.: Phys. Rev. *94*, 630 (1954)
5. Stejskal, E. O., Tanner, J. E.: J. Chem. Phys. *42*, 288 (1965)
6. Tanner, J. E.: Ph. D. Thesis (Chemistry) University of Wisconsin (1966)
7. Meiboom, S., Gill, D.: Rev. Sci. Instrum. *29*, 688 (1958)
8. Tanner, J. E.: J. Chem. Phys. *52*, 2523 (1970)
9. Packer, K. J., Rees, C., Tomlinson, D. J.: Molecul. Phys. *18*, 421 (1970)
10. Zupančič, I., Piiš, J.: J. Phys. (London) E *9*, 79 (1976)
11. Callaghan, P. T., Trotter C. M., Jolley, K. W.: J. Magn. Reson. *37*, 247 (1980)
12. Hrovat, M. I., Wade, C. G.: J. Magn. Reson. *44*, 62 (1981) and *45*, 67 (1981)
13. Cantor D. M., Jonas, J.: J. Magn. Reson. *28*, 157 (1977)
14. von Meerwall, E., Burgan, R. D., Ferguson, R. D.: J. Magn. Reson. *34*, 339 (1979)
15. Stilbs, P., Moseley, M. E.: Chem. Scripta *15*, 176 (1980)
16. Callaghan, P. T., Jolley K. W., Trotter, C. M.: JEOL News *16A*, 48 (1980)
17. McCall, D. W., Douglass, D. C., Anderson, E. W.: J. Polym. Sci. A *1*, 1709 (1963)
18. McCall, D. W., Huggins, C. M.: Appl. Phys. Letters *7*, 153 (1965)
19. von Meerwall, E.: J. Magn. Reson. *50*, 409 (1982)
20. Tanner, J. E.: J. Chem. Phys. 69, 1748 (1978)
21. Zientara, G. P., Freed, J. H.: J. Chem. Phys. *72*, 1285 (1980)
22. von Meerwall, E., Ferguson, R. D.: J. Chem. Phys. *74*, 6956 (1981)
23. von Meerwall, E.: Computer Phys. Commun. *17*, 309 (1979)
24. von Meerwall, E., Ferguson, R. D.: Computer Phys. Commun. *21*, 421 (1981)
25. Grosz, B., Kosfeld, R.: Messtechnik *7*, 171 (1969)
26. deGennes, P. G.: J. Chem. Phys. *55*, 572 (1971); also J. Phys. (Paris) *36*, 1199 (1975)
27. Doi, M., Edwards, S. F.: J. Chem. Soc. Faraday Trans II *74*, 1789, 1802, 1818, (1978); *75*, 38 (1979)
28. James, T. L., McDonald, G. G.: J. Magn. Reson. *11*, 58 (1973)
29. Boss, B. D., Stejskal, E. O., Ferry, J. D.: J. Phys. Chem. *71*, 1501 (1967)
30. von Meerwall, E., Ferguson, R. D.: J. Appl. Polym. Sci. *23*, 877 (1979)
31. Ferguson, R. D., von Meerwall, E.: J. Polym. Sci. Polym. Phys. Ed. *18*, 1285 (1980)

32. von Meerwall, E., Tomich, D. H., Hadjichristidis, N., Fetters, L. J.: Macromolecules *15*, 1157 (1982)
33. Cosgrove, T., Warren, R. F.: Polymer *18*, 255 (1977)
34. McCall, D. W., Douglass, D. C., Anderson, E. W.: J. Chem. Phys. *30*, 771 (1959)
35. Douglass, D. C. McCall, D. W.: J. Phys. Chem. *62*, 1102 (1958)
36. Hirschfelder, J. O., Stevenson, D., Eyring, H.: J. Chem. Phys. *5*, 896 (1937)
37. von Meerwall, E., Ferguson, R. D.: J. Chem. Phys. 72, 2861 (1980)
38. McCall, D. W., Anderson, E. W., Huggins, C. M.: J. Chem. Phys. *34*, 804 (1961)
39. Tanner, J. E.: Macromolecules *4*, 748 (1971)
40. Tanner, J. E., Liu, K.-J., Anderson, J. E.: Macromolecules *4*, 586 (1971)
41. von Meerwall, E., Grigsby, J., Tomich, D., Van Antwerp, R.: J. Polym. Sci. Polym. Phys. Ed. *20*, 1037 (1982)
42. Bueche, F.: Physical Properties of Polymers, Interscience, New York (1962)
43. Fujita, H.: Fortschr. Hochpolym.-Forsch. *3*, 1 (1961)
44. See, for example, Ferry, J. D.: Viscoelastic Properties of Polymers, 3rd ed. Wiley, New York (1980)
45. Flory, P. J.: Principles of Polymer Chemistry, Cornell U. Press, Ithaca, New York (1953), Ch. 14
46. Moseley, M. E.: Polymer Reports *21*, 1479 (1980)
47. Pimenov, G. G., Smechko, A. G., Azancheev, N. M., Skurda, V. D.: Akad. Nauk. USSR *20B*, 180 (1978)
48. Callaghan P. T., Pinder, D. N.: Macromolecules *13*, 1085 (1980)
49. Callaghan, P. T., Pinder, D. N.: Macromolecules *14*, 1334 (1981)
50. Pyun, C. W., Fixman, M.: J. Chem. Phys. *41*, 937 (1964)
51. Yamakawa, H.: J. Chem. Phys. *36*, 2995 (1962)
52. King, T. A., Knox, A., McAdam, J. D. G.: Polymer *14*, 293 (1973)
53. von Meerwall, E., Tomich, D. H., Grigsby, J., Pennisi, R., Fetters, L. J., Hadjichristidis, N.: Macromolecules (in press)
54. Callaghan, P. T., Pinder, D. N.: Polymer Bull. *5*, 305 (1981)
55. Brown, W., Stilbs, P., Johnsen, R. M.: J. Polym. Sci., Polym. Phys. Ed. *20*, 1771 (1982)
56. Kessler, D., Witte, H., Weiss, A.: Ber. Bunsenges, Phys. Chem. *73*, 368 (1969)
57. Woessner, D. E.: J. Phys. Chem. *67*, 1365 (1963)
58. Kosfeld, R., Goffloo, K.: Kolloid-Z. 247, 801 (1971)
59. Williams, M. L., Landel, R. F., Ferry, J. D.: J. Amer. Chem. Soc. *77*, 3701 (1955)
60. Maklakov, A. I., Smechko, A. G., Maklakov, A. A.: Macromolecular Sci. (Moscow) *19*, 2611 (1977)
61. Azancheev, N. M., Maklakov, A. I.: Macromolecular Sci. (Moscow) *21*, 1574 (1979)
62. Zupančič, I., Lahajnar, G., Blinc, R., Reneker, D. H., Peterlin, A.: J. Polym. Sci., Polym. Phys. Ed. *16*, 1399 (1978)
63. Moseley, M. E., Stilbs, P.: Chemica Scripta *16*, 114 (1980)
64. Nyström, B., Moseley, M. E., Stilbs, P., Roots, J.: Polymer *22*, 218 (1981)
65. von Meerwall, E., Ferguson, R. D.: J. Polym. Sci., Polym. Phys. Ed. *19*, 77 (1981)
66. von Meerwall, E., Van Antwerp, R.: Macromolecules *15*, 1115 (1982)
67. Vrentas, J. S., Duda, J. L.: J. Polym. Sci., Polym. Phys. Ed. *15*, 403 and 417 (1977); *17*, 1085 (1979)
68. Derbyshire, W., Duff, I. D.: Discuss Faraday Soc. *57*, 243 (1974)
69. Stilbs, P., Lindman, Bj.: J. Magn. Reson. *48*, 132 (1982)
70. von Meerwall, E., Ferguson, R. D.: J. Appl. Polymer Sci. *23*, 3657 (1979)
71. von Meerwall, E., Ferguson, R. D.: J. Chem. Phys. *75*, 937 (1981)
72. Muhr, A. H., Blanchard, J. M. V.: Polymer *23*, 1012 (1982) (Suppl.)
73. See, for example, "Diffusion in Polymers", J. Crank and G. S. Park (eds.) Academic Press, New York (1968)

Received February 14, 1983
J. D. Ferry (editor)

10 Appendix A
(added in proof)

Since the preparation of this review, several articles of importance to Sections 4 and 5 have appeared.

Fleischer [A1] reported on a PGSE study of linear polystyrene and linear and branched polyethylene melts ($M \leq 53000$) at 200 °C. A rapidly-attenuated echo component at low x (see eq. 2) was also [32] attributed to low-molecular-weight traces; some enhancement of D at the shortest t was thought to arise from gel-like fluctuations. The results follow eq. (7) with $n = -2$; since this slope holds below as well as above M_c, the reptation interpretation [26, 27] is in doubt.

PGSE measurements on poly(ethylene oxide) in aqueous dextran solutions were performed by Brown and Stilbs [A2] as function of the concentrations of both polymers. The results for D(PEO) depend on the product of the concentration and the intrinsic viscosity of the dextran (host) component, and suggest that coil overlap in the concentrated host solution is the principal impediment to PEO diffusion.

A comparison of PGSE [A3] and quasielastic light scattering measurements in semidilute poly(ethylene oxide) and polystyrene solutions was made by Brown, Johnson, and Stilbs [A4]. They concluded that the "slow" decay process observed in the light scattering experiments cannot, as had been assumed, be identified with center-of-mass self-diffusion (as unambiguously determined, e.g., by FGSE experiments), and that its origins are not yet fully understood.

Finally, two articles review recent experimental and theoretical developments with respect to self-diffusion in polymer solutions, either concentrated [A5], or dilute and semidilute [A6] with emphasis on scaling concepts. Both include references to the recent PGSE literature.

11 References (Appendix A)

A1. Fleischer, G., Polymer Bull. *9*, 152 (1983)
A2. Brown, W., Stilbs, P., Polymer *24*, 188 (1983)
A3. Stilbs, P., Colloid Interface Sci. *87*, 385 (1982)
A4. Brown, W., Johnsen, R. M., Stilbs, P., Polymer Bull. *9*, 305 (1983)
A5. Sundelöf, L.-O., Ber. Bunsenges. Physik. Chem. *83*, 329 (1983)
A6. Nyström, B., Roots, J., Progr. Polym. Sci. *8*, 333 (1982)

Photophysics of Excimer Formation in Aryl Vinyl Polymers

Steven N. Semerak and Curtis W. Frank
Department of Chemical Engineering, Stanford University, Stanford, CA 94305, USA

The objective of this review is to characterize the excimer formation and energy migration processes in aryl vinyl polymers sufficiently well that the excimer probe may be used quantitatively to study polymer structure. One such area of application in which some measure of success has already been achieved is in the analysis of the thermodynamics of multicomponent systems and the kinetics of phase separation. In the future, it is likely that the technique will also prove fruitful in the study of structural order in liquid crystalline polymers.

1 Introduction

1.1 Scope of the Review

Fluorescence techniques have been demonstrated in recent reviews [1] to be powerful methods for obtaining detailed information on the molecular structure of biopolymers and synthetic polymers. The objective of the present review is to concentrate on two aspects of the photophysics of synthetic polymers — excimer formation and singlet exciton migration. Both topics have been considered recently [1] but in less detail.

Our motivation for offering a further consideration of excimer fluorescence is that it is a significant feature of the luminescence behavior of virtually all aryl vinyl polymers. Although early research was almost entirely devoted to understanding the intrinsic properties of the excimer complex, more recent efforts have been directed at application of the phenomenon to solution of problems in polymer physics and chemistry. Thus, it seems an appropriate time to evaluate existing information about the photophysical processes and structural considerations which may influence excimer formation and stability. This should help clarify both the power and limitations of the excimer as a molecular probe of polymer structure and dynamics.

One of the ambiguities associated with interpretation of the photophysics of the aryl vinyl polymers and copolymer variants is the contribution due to singlet exciton migration. In many cases it is difficult to differentiate between an explanation of a fluorescence observation that is based on a decrease in the number of traps consisting of excimer forming sites (EFS) and an alternative explanation based upon a decrease in the energy migration efficiency. Unfortunately, this question cannot be resolved in this review since there is, as yet, no published theoretical model which adequately treats the problem of exciton migration among and between polymer chains having high chromophore densities.

One point which can be settled qualitatively, however, is whether singlet exciton migration does in fact occur in the aryl vinyl polymers. It will be shown that available evidence supports energy migration as an important feature of the photophysics of polystyrene (PS), poly(1-vinyl naphthalene) (P1VN), and poly(2-vinyl naphthalene) (P2VN), the homopolymers which are the subject of the majority of the review.

In the first part of the review, we will cover the major photophysical species and processes occurring in PS, P1VN, and P2VN under the experimental conditions applicable to solid blends. Such systems have been the subject of a series of studies in this laboratory. [2] Special attention will be given to the effects of oxygen in the photophysics, and to the effects of the rigidity of the host matrix relative to solution. Because the number of EFS is a major factor in determining the ratio of excimer (I_D) to monomer (I_M) fluorescence intensities, the molecular structure of PS and the conformationally similar P2VN will be discussed and the number of intramolecular EFS will be estimated. In addition, the photobehavior of the naphthalenophanes will be reviewed. The structure of these bichromophoric compounds can be altered to produce ring pairs held at various distances and angles. Thus, the range of permissible geometries for an EFS can be determined. In the second part of the literature review, evidence for and against migration of singlet state energy in aryl vinyl polymers will be presented. The most likely mechanism of the energy migration will be discussed, since an a priori estimation of the rate of energy migration is highly desirable.

1.2 Photophysical Nomenclature for Rigid Systems

The most widely used nomenclature in the field of luminescence of aromatic molecules is that proposed by Birks. [3] The photophysical species and processes in this scheme which are encountered in rigid systems of aryl vinyl polymers are described in Tables 1–4. Triplet excimers have been omitted since it will be shown later that all triplet states play minor roles in the room-temperature, air saturated P2VN blends studied recently [2].

In all of the processes described in Table 4, only the excited state energy migrates, since translational diffusion of the chromophores is prohibited by the rigid environment. The sole exception is the possible diffusion of oxygen through the glassy polymer in its quenching ($k'_{QT}[^3Q]$) of the long-lived triplet state. The spatial relationship of the chromophores involved in the migration and transfer processes has been left unspecified for the present. The process may involve adjacent or nonadjacent intramolecular groups, or it may involve intermolecular groups.

2 Photophysical Species and Monochromophoric Processes in Aryl Vinyl Polymers

The photoprocesses covered in Table 3, which do not require energy migration, will be reviewed in four parts: UV absorption, fluorescence (non-excimeric), triplet species, and excimers. Each part will be structured as follows. First, a list of questions will be presented which will provide a focus for the literature search. Second, any basic theory which addresses the search questions will be given. Third, a review of experimental results will be made. These results will be organized by compound, in order of increasing molecular weight. In general, only compounds containing the phenyl or naphthyl chromophore will be reviewed. We will, however, comment on results for other aryl vinyl polymers when comparable information is unavailable for the phenyl or naphthyl systems. Finally, we will attempt to answer the search questions posed at the beginning of each review part.

Table 1. Intrinsic Photophysical Species in Aryl Vinyl Polymers in a Rigid Environment

Symbol	Description
M	Chromophore or "monomer" in ground singlet state
M^*	Monomer in 1st excited singlet state
$^3M^*$	Monomer in 1st excited triplet state
M–M	Excimer-forming site (EFS) in ground singlet state
M^*–M or D^*	Excimer in 1st excited singlet state
M–($^3M^*$)	EFS with one chromophore in 1st excited triplet state

Table 2. Quenching and Impurity Species Present in Aryl Vinyl Polymers in a Rigid Environment

Symbol	Description
^{3}(M–Q)	Quenched monomer site in ground triplet state ($Q = O_2$)
^{3}(M*–Q) or ^{3}E*	Quencher exciplex in 1st excited triplet state
M–(Q*)	Quenched monomer site with quencher in 1st excited singlet state ($Q^* = {}^1O_2^*$)
^{3}Q	Free quencher which may diffuse within the $^3M^*$ lifetime ($Q = O_2$)
Y	Impurity chromophore in ground singlet state
Y*	Impurity chromophore in 1st excited singlet state, which lies at lower energy than M*
^{3}Y*	Impurity chromophore in 1st excited triplet state

Table 3. Monochromophoric Photophysical Processes in Aryl Vinyl Polymers in a Rigid Environment

Process	Description	Rate, s^{-1}
$M + h\nu_{ex} \rightarrow M^*$	Absorption by monomer	—
$M\text{–}M + h\nu_{ex} \rightarrow D^*$	Direct excitation of an EFS	—
$^3(M\text{–}Q) + h\nu_{ex} \rightarrow {}^3E^*$	Direct excitation of quenched monomer ($Q = O_2$)	—
$M^* \rightarrow M + h\nu_M$	Monomer fluorescence	k_{FM}
$M^* \rightarrow M$	Monomer internal conversion	k_{GM}
$M^* \rightarrow {}^3M^*$	Monomer intersystem crossing	k_{TM}
$D^* \rightarrow M\text{–}M + h\nu_D$	Excimer fluorescence	k_{FD}
$D^* \rightarrow M\text{–}M$	Excimer internal conversion	k_{GD}
$D^* \rightarrow M\text{–}({}^3M^*)$	Excimer intersystem crossing	k_{TD}
$^3M^* \rightarrow M + h\nu_T$	Monomer phosphorescence	k_{PT}
$^3M^* \rightarrow M$	Intersystem crossing to ground	k_{GT}
$^3E^* \rightarrow {}^3(M\text{–}Q)$	Triplet quencher exciplex internal conversion	k_{GX}
$Y + h\nu_{ex} \rightarrow Y^*$	Absorption by impurity	—
$Y^* \rightarrow Y + h\nu_Y$	Impurity fluorescence	k_{FY}
$Y^* \rightarrow Y$	Impurity internal conversion	k_{GY}
$Y^* \rightarrow {}^3Y^*$	Impurity intersystem crossing	k_{ZY}
$^3Y^* \rightarrow Y + h\nu_Z$	Impurity phosphorescence	k_{PZ}
$^3Y^* \rightarrow Y$	Intersystem crossing to ground	k_{GZ}

Table 4. Bichromophoric Processes in Aryl Vinyl Polymers in a Rigid Environment

Process	Description	Rate, s^{-1}
$M^* + M \rightarrow M + M^*$	Singlet energy migration	k_{MM}
$^3M^* + M \rightarrow M + {}^3M^*$	Triplet energy migration	k_{TT}
$M^* + M\ M \rightarrow M + D^*$	Migrative excitation of an EFS	k_{DM}
$D^* + M \rightarrow M\text{–}M + M^*$	Excimer dissociation to excited nearest-neighbor monomer	k_{MD}
$^3M^* + {}^3M^* \rightarrow M^* + M$	Migrative triplet annihilation	$k_{MTT}\,[{}^3M^*]$
$M^* + {}^3(M\text{–}Q) \rightarrow M + {}^3E^*$	Migrative quenching of singlet monomer	k_{XM}
$^3M^* + {}^3(M\text{–}Q) \rightarrow M + M\text{–}(Q^*)$	Migrative quenching of triplet monomer ($Q^* = {}^1O_2^*$)	k_{QT}
$^3M^* + {}^3Q \rightarrow M + Q^*$	Diffusive quenching of triplet monomer ($Q = O_2$)	$k'_{QT}\,[{}^3Q]$
$M^* + Y \rightarrow M + Y^*$	Singlet energy transfer to impurity	k_{YM}
$^3M^* + Y \rightarrow M + {}^3Y^*$	Triplet energy transfer to impurity	k_{ZM}

2.1 Absorbing Species

2.1.1 Introduction

The major questions of this section are:
- What are the absorbing species in the aryl vinyl polymers?
- How does the UV absorbance of an aryl vinyl polymer differ from that of the matching monochromophoric compound, given that the polymer is free from defects and impurities?

If any ground-state interactions occur among the chromophores of the polymer, the UV absorbance will be altered. However, such alterations are more likely to be caused by impurity chromophores on the polymer chain. Since it is extremely difficult to obtain high polymer which is free from defects and impurities, another standard for the polymer UV absorbance is desired.

Bichromophoric model compounds can provide such standards, since these compounds can be highly purified and characterized by techniques inapplicable to polymers. For example, the racemic compound 2,4-diphenylpentane models the syndiotactic dyad of PS, while the *meso* 2,4-diphenylpentane models the isotactic dyad of PS. The 2,4-diarylpentanes are the best models of aryl vinyl polymers, since the populations of the rotational conformers of the model compounds are similar to those of the dyads of the polymer.

2.1.2 Bichromophoric Model Compounds Containing Phenyl or Naphthyl Rings

The first comprehensive study of diphenylalkanes was reported by Hirayama [4], who studied all α,ω-diphenylalkanes from diphenylmethane to 1,6-diphenylhexane, various compounds related to 1,3-diphenylpropane, and 1,3,5-triphenylpentane. It was noted that the latter compound and all 1,3-diphenylalkanes had "almost the same" UV absorbance spectrum as ethyl benzene. Similar conclusions were reached by Rice and co-workers [5] for 1,2-diphenylethane, 1,3-diphenylpropane, and 1,4-diphenylbutane. A more detailed study was made by Monnerie and co-workers [6a] on *meso*- and *dl*-2,4-diphenylpentane and on syndiotactic, isotactic, and heterotactic 2,4,6-triphenylheptane. They found that the maximum UV absorbances for the *dl* and syndiotactic compounds were about 5% larger than for the *meso* and isotactic compounds, respectively. In a separate study by Monnerie et al., [6b] the peak and shoulder wavelengths in the absorbance spectra of *dl* and *meso*-2,4-diphenylpentane were matched most closely by those in the spectrum of isopropylbenzene relative to other alkylbenzenes. Also, the absorbance of the *dl* compound was again found to be from 0 to 15% larger than the absorbance of the *meso* compound at $\lambda = 259$ and 268 nm, respectively. These UV absorbance spectra generally agree with others published [7,8] for 2,4-diphenylpentane.

There are fewer UV absorbance studies of dinaphthylalkanes. Chandross and Dempster [9] have studied 1,2-bis(1-naphthyl)ethane, 1,3-bis(1-naphthyl)propane and 1,4-bis(1-naphthyl)butane, as well as 1,3-bis(2-naphthyl)propane and the compound 1-(1-naphthyl)-3(2-naphthyl)propane. The latter had the same absorbance spectrum as a 50/50 mixture of 1- and 2-methylnaphthalene, while the bis compounds were shown to have the same absorbance spectrum as the corresponding methylnaphthalene isomer. These studies were made in a 90/10 v/v mixture of methylcyclohexane/isopentane at 77 and 300 K. Since the stability of the model compound is an important consideration, we note that, of all of the above compounds, a slow photochemical reaction was reported [10] only for 1,3-bis(1-naphthyl)-propane.

De Schryver and co-workers [11] have confirmed Chandross' result for the UV absorbance of 1,3-bis(2-naphthyl)propane. Nishijima et al. [12] have stated that the absorbance spectrum of *meso*- and dl-2,4-bis(2-naphthyl)pentane and of the compounds 1,3-bis(2-naphthyl)A, where A = propane, butane, pentadecane, and 5-phenylpentane, is similar to the absorbance spectrum of 2-ethylnaphthalene. Finally, an unusual result has been obtained by De Schryver et al. [13] for the compound bis(1-(2-naphthyl)ethyl)ether. The *meso* compound gave a lower value of I_D/I_M, the ratio of excimer to monomer fluorescence intensities, under excitation at 304 nm relative to excitation at 285 nm, while the *dl* compound had no such excitation dependence. The UV absorbance spectra of these compounds were not reported, however.

In short, the UV absorbance spectra of diphenyl and dinaphthylalkanes are substantially the same as that of the corresponding alkylarene. There are no major wavelength or intensity shifts of the spectral bands, and no signs of any additional bands at lower energy. Since there are no ground-state dimers present in these model compounds that are detectable by UV absorption, we expect that no such dimers will be found for PS or P2VN in dilute solution.

2.1.3 Polystyrene

2.1.3.1 Comparison with Alkyl Benzene Model Compounds

It has been known for quite some time that the extinction coefficient of PS at its absorption maximum is roughly independent of the molecular weight of the polymer. [14] Some of the pre-1960 reports may be inaccurate due to the incomplete removal of styrene from the polymer samples, however.

Several reports have appeared in which the UV absorption spectrum of PS was found to be qualitatively similar to an alkyl benzene model compound. Vala and Rice [15] demonstrated the similarity between atactic PS and ethyl benzene; this was confirmed for atactic PS in cyclohexane by Hirayama [16] and in tetrahydrofuran by Abuin [17]. Bovey and co-workers [20] found that absorbance spectra for isotactic PS, atactic PS, and styrene-methylmethacrylate copolymers in 1,2-dichloroethane were roughly similar to that of toluene.

More quantitative studies have shown that, although some hypochromism effects exist, they are small. For example, Monnerie [19] examined the integrated molar extinction coefficient of atactic PS in chloroform and found only a 2% decrease for the polymer relative to ethyl benzene. In addition, Vala and Rice [15] reported a 10% decrease in absorption for the 260 nm band of isotactic PS relative to atactic PS, which was qualitatively confirmed by Longworth [18]. Similarly, Cantow [8] observed a 4% hypochromism of the 261.5 nm band of isotactic PS in dioxane relative to atactic PS. Finally, Cantow [8] observed strong hypochromism (relative to atactic PS) ranging from 19% at 262 nm to 32% at 269 nm for an alternating styrene-methyl methacrylate copolymer and for random copolymers having a low styrene content.

In summary, the UV absorbance spectrum of atactic PS in both poor and good solvents was found to be substantially unchanged from that of ethyl benzene or isopropylbenzene. The hypochromism at 261 nm of isotactic PS relative to atactic

PS, and of *meso*-2,4-diphenylpentane relative to the *dl* compound suggests that the hypochromic effect does not require a large number (ca. 10) of interacting rings, as proposed by Vala and Rice [15].

2.1.3.2 Identification of In-Chain Impurities

The UV absorbance and other photophysical properties of PS in bulk, recently reviewed by McKeller and Allen [21], have been studied by Klöpffer [22] at wavelengths above 290 nm, where the phenyl chromophore does not absorb. Despite careful removal of styrene by polymer precipitation or by gel permeation chromatography, PS samples from anionic and radicalic polymerizations were found to have $\varepsilon_{290}/\varepsilon_{260} = 10^{-3}$. The same ratio for isopropylbenzene, by contrast, is less than 2×10^{-4}. Based on this and on the luminescence of PS samples excited at wavelengths above 290 nm, Klöpffer suggested that the following chromophores (R = phenyl) were present as in-chain impurities in PS:

$$-HC{=}CHR,\quad -H_2C-\underset{\displaystyle R}{\underset{|}{C}}{=}\overset{\displaystyle R}{\overset{|}{C}}-CH_2-,\quad \text{and}\quad -H_2C-COR\,.$$

The presence of these impurities at 100–1,000 ppm could account for the UV absorbance at 290 nm and above. The formation of oxygen-phenyl group complexes [23,24] has also been suggested [22,36] to be partly responsible for the PS absorption tail above 290 nm. Finally, the $-H_2C-\underset{\displaystyle R}{\underset{|}{C}}{=}\overset{\displaystyle R}{\overset{|}{C}}-CH_2-$ linkage was detected by Soutar and co-workers [25] in their study of head-to-head PS in films and in solution.

2.1.4 Poly(1-vinyl naphthalene) and Poly(2-vinyl naphthalene)

2.1.4.1 Comparison with Alkyl Naphthalene Model Compounds

Because P1VN and P2VN do not have the same commercial importance as PS, less data are available for the UV absorbance of the vinylnaphthalene polymers and copolymers. Laitinen and co-workers [26] were the first to report the absorbance spectrum between 240 and 300 nm of P2VN in chloroform solution. They found $\varepsilon_{max} = 4{,}100$ at $\lambda_{max} = 278$ nm. Apparently, Vala et al. [5] were the first to note that P1VN and 1-substituted "naphthalenes" had similar spectra. More recently, Nishijima [27] published spectra of P1VN and 1-ethyl naphthalene and reported $\varepsilon_{max} = 5{,}500$ at $\lambda_{max} = 288$ nm for P1VN in dichloromethane. The same group later published [28] a spectrum of P2VN which confirmed Laitinen's result and also showed the high energy bands above 300 nm that are characteristic of 2-substituted naphthalenes. Note that the absorbance maxima of 2-isopropylnaphthalene [29] and 1-isopropylnaphthalene [30] appear at 274 nm and 285 nm, respectively, and that

$\varepsilon_{max} = 5{,}100$ and 6,650, respectively. Thus, the absorbance spectra of both P1VN and P2VN are red-shifted about 4 nm and show 15% less absorbance at the maximum, relative to the isopropylnaphthalenes.

While other researchers have commented on the UV absorbance of vinylnaphthalene polymers and copolymers, the spectra are generally not shown or tabulated. De Schryver and co-workers [11,13] have reported that the absorption spectrum of P2VN is experimentally superimposable on that of 2-methylnaphthalene. They also showed that the excitation spectrum of P2VN analyzed at 420 nm is identical to the absorption spectrum, but that the excitation spectrum analyzed at 335 nm is enhanced between 280 and 320 nm. In addition, they observed a doubling of the P2VN fluorescence ratio for excitation at 285 nm relative to 304 nm. This doubling of I_D/I_M was also seen for the model compound *meso*-bis(1-(2-naphthyl)-ethyl) ether [13]. Fox and co-workers [31] found no anomalous UV absorption in P1VN, P2VN or 1-vinylnaphthalene/styrene copolymers. David et al. [32] have stated that the absorbance spectrum of 1-vinylnaphthalene/methyl methacrylate copolymers does not depend on copolymer composition, in contrast to the behavior seen [8] for styrene/methyl methacrylate copolymers.

2.1.4.2 *Identification of In-Chain Impurities*

Although most studies indicate that P1VN and P2VN do not absorb at wavelengths longer than 330 nm, there are several reports to the contrary. Irie and co-workers [33] found new absorption bands at 331 and 339 nm for a low molecular-weight (about 3,000) P1VN sample in cyclohexane. These bands also appeared in CCl_4 and *n*-hexane solution, though not as strongly as in cyclohexane, and were completely absent in dichloromethane or tetrahydrofuran. The fluorescence spectrum of P1VN in dilute cyclohexane solution was structured, with peaks at 342, 359, 378, and 397 nm, while in dichloromethane a broad peak appeared at the usual excimer position of 400 nm, with shoulders at 340 and 360 nm. Anomalous absorbance and fluorescence of a low molecular-weight P2VN sample was also observed [33] to be quite similar to that of P1VN.

The results of Irie et al. [33] have not been confirmed nor directly refuted in other studies of P1VN and P2VN, since cyclohexane or *n*-hexane are nonsolvents for these polymers when the molecular weight is greater than about 5,000. However, Fox et al. [31] examined the fluorescence of the fraction of a P1VN sample that was soluble in a 99:0.5:0.5 mixture of cyclohexane, tetrahydrofuran and ether. They found that the fluorescence spectrum of the P1VN fraction was identical (except for a decrease in I_D/I_M) to the spectrum of the original P1VN sample in a 50:50 mixture of tetrahydrofuran and ether.

Pratte and Webber [34] have recently scrutinized the absorbance and emission of several P2VN samples with molecular weight greater than 13,000. They found significant (though structureless) UV absorption above 320 nm that was strongest for the lowest molecular-weight sample, and was observable in dichloromethane or benzene solution. Also, the fluorescence spectrum of the low molecular-weight P2VN sample in benzene excited at 338 nm showed a maximum at 370 nm and shoulders at about 350 and 385 nm. Excitation at 295 nm produced a fluorescence spectrum with the usual 410 nm excimer maximum, without shoulders near 360 and 380 nm.

Although Irie et al. [33] performed a hydrogenation of their P1VN and P2VN samples in an attempt to remove olefinic unsaturation, their assignment of the new absorption and fluorescence bands as due to a groundstate dimer in these polymers seems unfounded. For P2VN, Pratte and Webber [34] attributed the new absorption bands to a moiety with vinylnaphthalene conjugation. By utilizing more vigorous hydrogenation conditions, they found that the absorbance of P2VN at 340 nm decreased by 25% after one hydrogenation step. Also, the increase of ε_{340} with decreasing molecular weight [34] suggests a chain-end impurity formed during the P2VN synthesis. Finally, there have been no signs of the new P2VN absorption and fluorescence bands in any of the studies of bis (2-naphthyl) compounds presented earlier or to be presented in the next section. The impurity hypothesis successfully explains all these points, except the observed increase in ε_{340} for P2VN in cyclohexane relative to dichloromethane [33]. Future studies of model compounds such as $CHR{=}CH{-}CHR{-}CH_3$, where R = 1- and 2-naphthyl, may help explain why there is a solvent dependence of the absorption and emission spectra of the polyvinylnaphthalene samples of Irie et al.

The case for the impurity hypothesis is not so strong for P1VN as for P2VN, however. First, the absorption spectrum of 1-propenylnaphthalene [35] is structureless for $\lambda > 320$ nm; Irie et al. [33] saw bands at 330 and 339 nm for P1VN. Second, the fluorescence spectrum of 1,3-bis(1-naphthyl)propane exhibits new bands [33] at 343 and 358 nm for solid cyclohexane solutions at 77 K; these bands are absent at room temperature for glassy methylcyclohexane/isopentane solutions [9] at 77 K. More work is necessary to determine the effect of hydrogenation on the absorbance and emission behavior of P1VN, and the unique role of cyclohexane in promoting the anomalous behavior of P1VN.

2.2 Non-Excimeric Fluorescing Species

2.2.1 Introduction

Emission from M*, monomer fluorescence, is the topic of this section. The following questions will be considered:

- How is the monomer fluorescence of aryl vinyl polymers or intramolecular excimer-forming compounds distinguished from that of monochromophoric compounds?
- What are the effects of a rigid polymer matrix and of oxygen on the fluorescence lifetime and quantum yield?

The lifetime of monomer fluorescence in the absence of routes to excimer formation is $\tau_M = (k_{FM} + k_{GM} + k_{TM})^{-1} = k_M^{-1}$, and the intrinsic quantum yield is $Q_M = k_{FM}/k_M$. The lifetime and intrinsic quantum yield of excimer fluorescence, τ_D and Q_D, will be considered in a later section.

2.2.2 Comparison of Bichromophoric and Monochromophoric Compounds

The first question cannot be answered directly even for a bichromophoric compound capable of forming intramolecular excimers. The simplest kinetic scheme in fluid solution involves the six rates k_{FM}, k_M, k_{FD}, k_D, k'_{DM}, and k'_{MD}. The latter two primed rates refer to excimer formation and dissociation induced by rotational sampling (in contrast to the processes occurring through energy migration in rigid solution). However, only four quantities can be determined experimentally, namely the net quantum yield of excimer and monomer emission φ_D and φ_M, and the two decay rates λ_1 and λ_2 [71]. The standard practice in the literature is to determine k_{FM} and k_M for the corresponding monochromophoric compound and assume that these rates are applicable to the bichromophoric compound. [37] In this review section, we will examine bichromophoric compounds that do not form excimers, yet which may be capable of other intramolecular quenching processes.

One method of directly determining Q_M and Q_D for excimer-forming compounds involves the relationship $\varphi_D/Q_D + \varphi_M/Q_M = 1$. If φ_M and φ_D are measured in a series of solvents, and if Q_M and Q_D are assumed to be independent of solvent, then Q_M and Q_D can be obtained from extrapolation of the φ_D *vs.* φ_M relation to the x- and y-axis intercepts. A second direct method requires a number of samples of an aryl vinyl polymer with differing molecular weights. If Q_M and Q_D are assumed to be independent of molecular weight, then they can again be obtained from the φ_D *vs.* φ_M relation for all of the polymer samples. The rate k_M can be determined if k_{FM} is assumed to be identical to that of the monochromophoric compound. It follows that $k_M = k_{FM}/Q_M$. While these methods are rarely used in the literature, it is worthwhile to review the assumptions that Q_M and Q_D are independent of solvent, or of the molecular weight and structure of the compound.

The monomer fluorescence of compounds having the structure $R(CH_2)_xR$ is quite similar to that of the corresponding monochromophoric compound. When R = phenyl and x = 1, 2 or 4–6, the fluorescence spectra and monomer quantum yields of these compounds in deoxygenated cyclohexane were found [4, 37] to be within 12% of $\varphi_M = 0.18$ [37] for toluene. For x = 3, the monomer quantum yield was reduced by a small amount of intramolecular excimer formation [38]. The fluorescence spectra [9] and monomer quantum yields [39] of compounds having R = 1-naphthyl and x = 1, 2 or 4, or the compounds $R\text{-}CH_2OCH_2CH_2\text{-}R$ and $R\text{-}CH_2OCH_2CH_2OCH_2\text{-}R$, were found to be within 10% of $\varphi_M = 0.21$ or 0.10 for 1-methylnaphthalene and methyl 1-naphthylmethyl ether, respectively. However, when x = 4, weak excimer fluorescence was observed for solutions near T = 223 K [9]. When R = 2-naphthyl and x = 5, 7 or 12, the fluorescence spectra and monomer quantum yields of these compounds in degassed tetrahydrofuran were found [40] to be within 6% of $\varphi_M = 0.24$ [40] for 2-ethylnaphthalene at 298 K, and the similarity was also found at T = 173 K up to room temperature. The same was true for the compound 1,5-*bis*(2-naphthyl)-3-phenylpentane [41]. For compounds having R = 2-naphthyl and x = 1, 2, 4 or 6, or the compounds $RCH_2(OCH_2CH_2)_nOCH_2R$ having n = 1 or 2, the monomer quantum yields in degassed cyclohexane were found [29] to be within 25% of $\varphi_M - 0.27$ [12] or 0.11 for 2-methylnaphthalene and methyl 2-naphthylmethyl ether, respectively. Interestingly, φ_M for the compound $R\text{-}CH_2OCH_2CH_2\text{-}R$ was reduced by a small amount of intramolecular excimer formation [39] at 293 K, and a similar observation was made for the compound (1-naphthyl)-$(CH_2)_3$-(2-naphthyl) for solutions near T = 223 K [9].

In all the above studies [4, 9, 37–41] the 1,3-diarylpropanes had a reduced monomer quantum yield due to substantial intramolecular excimer formation, but the monomer fluorescence spectrum was unchanged from that of the analogous monochromophoric molecule. Outside of excimer formation, these studies show that there are no intramolecular processes that alter the decay of the phenyl or naphthyl chromophores.

2.2.3 Intramolecular Quenching by Carbonyl

A notable exception may be drawn from the fluorescence behavior of poly(phenylalkyl methacrylate)s [43] and of copolymers of methyl methacrylate [44] or methyl acrylate [45] containing less than 0.5 mole percent 1-vinylnaphthalene comonomer. Intramolecular excimer formation was not observed for any of these polymers, facilitating the study of monomer fluorescence. Abuin and co-workers [43] found that the monomer fluorescence lifetimes and quantum yields of poly(benzyl methacrylate), poly(2-phenylethyl methacrylate), and poly(3-phenylpropyl methacrylate) were 20–55% less than the respective model compounds benzyl acetate, 2-phenylethyl acetate, and 3-phenylpropyl acetate. The decrease was attributed to intramolecular quenching of the phenyl group by the carbonyl moiety. Indeed, the lifetime and quantum yield of ethylbenzene were found [43] to be 45% less when determined in degassed ethyl acetate solution relative to cyclohexane solution, presumably caused by the analogous intermolecular quenching process.

A similar but smaller intramolecular quenching effect was seen by Phillips and co-workers [44,45] for 1-vinylnaphthalene copolymers incapable of excimer fluorescence. The monomer fluorescence lifetime of the 1-naphthyl group in the methyl methacrylate copolymer [44] was 20% less than the lifetime of 1-methylnaphthalene in the same solvent, tetrahydrofuran. However, no difference in lifetimes was observed between the 1-vinylnaphthalene/methyl acrylate copolymer [45] and 1-methylnaphthalene. To summarize, the nonradiative decay rate of excited singlet monomer in polymers, $k_{GM} + k_{TM}$, may not be identical to that of a monochromophoric model compound, especially when the polymer contains quenching moieties and the solvent is fluid enough to allow rapid intramolecular quenching to occur.

2.2.4 Influence of Matrix Rigidity and Dissolved Oxygen

The question remains whether there are significant differences in τ_M and Q_M for phenyl and naphthyl chromophores in rigid solvent matrices at room temperature relative to fluid solution. The data of Jones and Calloway [46] on the fluorescence lifetimes of naphthalene and perdeuteronaphthalene in degassed samples of PS, PMMA, and various fluid solvents at 298 K show that there is no difference (within 10%) for τ_M measured in polymeric or fluid solvents, and that there is no difference between the PS and PMMA hosts. While values of Q_M are not given for these systems [46], the tabulation by Birks [42] for naphthalene in various fluid solvents shows that there is no difference (within 10%) for Q_M. However, Q_M given for naphthalene in 95:5 ethanol:water solutions is exceptionally low. Otherwise, τ_M given by the two sources [42,46] for naphthalene agree well, and an earlier report [47] on τ_M of naphthalene and perdeuteronaphthalene in PMMA is consistent with the data of Jones and Calloway. Assuming that k_{FM} is identical for naphthalene in the PS and PMMA hosts, we conclude from the constancy of τ_M that Q_M is the same in both hosts.

Finally, the question of whether the oxygen present in air-saturated PS and PMMA hosts can quench fluorescence is considered. In a study of naphthalene in a PS host, Nowakowska et al. [36] have shown that the increase in k_M for air-saturated films relative to degassed films is less than 5%. Offen and co-workers [48] have studied pyrene in PMMA and PS hosts. They found that degassed and air-saturated films gave the same value of τ_M for pyrene, to within 5%. Moreover, τ_M was independent of the polymer host.

2.3 Triplet Species

2.3.1 Introduction

Phosphorescence and delayed monomer and excimer fluorescence have been observed in P1VN [50-54] and P2VN [55-58] in dilute glassy solution and in neat films at 77 K. The basic mechanism for the delayed fluorescence for P1VN and P2VN is considered to be triplet formation $^3M^*$ through monomer (k_{TM}) or excimer (k_{TD}) intersystem crossing, followed by migrative annihilation of two triplets to form one excited singlet (k_{MTT}). Competing with triplet-triplet annihilation is the normal process of triplet decay (k_T), comprised of intersystem crossing to ground (k_{GT}) and phosphorescence (k_{PT}). Triplet processes in P1VN and P2VN are important because triplet formation from excited singlet states is the major decay process in naphthalene at room temperature [65] ($k_{TM}/k_M = 0.8$).

The questions to be considered in this section are:

- How much does triplet-triplet annihilation, k_{MTT}, enhance the fluorescence yield of aryl vinyl polymers, relative to the fluorescence yield of the analogous monochromophoric compounds?
- How does the increase in temperature from 77 K to room temperature affect k_{MTT} and k_T?
- How does the presence of oxygen affect k_{MTT} and k_T?

2.3.2 Comparison of Polymeric and Model Compounds

Because of the difficulty and complexity of determining the delayed fluorescence yield, all the reports in the literature consider only the decay behavior of the delayed luminescence of P1VN [50,51] and P2VN. [56,58] Both the phosphorescence and delayed fluorescence decay nonexponentially at short times. The long-time phosphorescence intensity of the polymer decreases exponentially ($k_T \sim 0.6\ s^{-1}$), at a rate about 50% faster than that of the corresponding ethylnaphthalene compound under the same conditions. Because triplet-triplet annihilation must occur via triplet energy migration in these glassy solutions at 77 K, representation of the annihilation rate by $k_{MTT}[^3M^*]^2$ is a gross oversimplification. Nevertheless, we conclude from the initial decay rates of phosphorescence and delayed fluorescence in the polymers at 77 K that $k_{MTT}[^3M^*]$ could be 10–1000 times as large as k_T.

Studies of the triplet-triplet absorption spectrum of P1VN [59] and P2VN [60,61] in degassed fluid solution have shown that triplet annihilation is still important at room temperature. The triplet absorbance of dilute poly(vinylnaphthalene) solutions is only 0.05–0.2 times as large as that of the corresponding ethylnaphthalene solution having the same concentration of naphthyl groups. While quantitative data on k_{MTT} at room temperature are unavailable, the ordinary triplet decay rate k_T has been extensively studied for monochromophoric compounds.

2.3.3 Influence of Matrix Rigidity and Dissolved Oxygen

In contrast to the singlet decay rate k_M, k_T is quite sensitive to the viscosity of the host matrix. For example, k_T for 2-ethylnaphthalene goes from 0.4 s^{-1} [58] in 2-methyltetrahydrofuran glass at 77 K, to 100 s^{-1} [60] in degassed solution at 298 K. However, k_T for naphthalene in degassed PMMA films [62] at 300 K is only 0.56 s^{-1}, marginally greater than the 77 K value of 0.41 s^{-1}. It seems likely that k_{MTT} will also remain roughly independent of temperature when determined in glassy polymer hosts like PMMA. A report of delayed fluorescence from a 1 g/l solution of naphthalene in a degassed PMMA film at room temperature [63] indicates that k_{MTT} does not decrease substantially as the temperature is raised.

The last question to be considered is the effect of oxygen on the lifetime of the triplet state in rigid solution. Although the singlet state in room-temperature polymer hosts is scarcely affected by O_2, the triplet state should be much more susceptible to quenching because of its much longer lifetime. This was first demonstrated by Oster and co-workers [64] and later verified by numerous studies [65]. Most of the quenching at room temperature can be attributed to diffusional quenching $k'_{QT}[O_2]$ in the glassy polymer hosts, since the phosphorescence of oxygen-saturated samples can be restored by cooling to 77 K. Oxygen virtually eliminates the room-temperature phosphorescence of naphthalene in PMMA, as shown by Jones and Siegel [62]. Extrapolation of their data gives $k_T = 1{,}300\ s^{-1}$) for air-equilibrated samples, more than 2000 times greater than the degassed value of 0.56 s^{-1}. It seems likely that triplet-triplet annihilation in aryl vinyl polymer samples exposed to the air would be insignificant, especially since the annihilation rate is proportional to $[^3M^*]^2$.

In conclusion, the enormous triplet quenching rate in air-equilibrated polymer hosts at room temperature dominates all other triplet processes. Triplet-triplet annihilation in P1VN and P2VN under such conditions is assumed to yield a negligible amount of delayed fluorescence. The intersystem crossing of excited singlet states to triplets is, in effect, a radiationless singlet decay process in P1VN and P2VN.

2.4 Structure of Excimers and Excimer-Forming Sites

2.4.1 Introduction

A major problem in modeling the photophysics of aryl vinyl polymers is to identify and characterize both the intramolecular excimers made possible by the flexible hydrocarbon linkages between chromophores, as well as the intermolecular excimers occurring when the segment density is high. In rigid host matrices, excimers must occur at ring pairs whose conformation is sufficiently close to that of the excimer such that electronic excitation of one of the rings leads immediately to an excimer [66]. At the most, a chromophore could diffuse only 1 Å during its lifetime, assuming the same diffusion constant of 10^{-9} cm^2/sec as observed for oxygen in PMMA at room temperature [67].

The literature review on excimers will be directed along the two rather interdependent lines of structure and properties. In this section, the structure and symmetry of naphthalene excimers will be considered, and the number of intramolecular EFS allowed by the rotational isomers of an aryl vinyl polymer will be determined. In the following section, the lifetime τ_D and intrinsic quantum yield Q_D of excimer fluorescence from intra- and intermolecular naphthalene excimers will be examined.

The questions to be considered first are:

- How many stable excimer structures exist for naphthalene and its alkyl-substituted derivatives?
- Does the UV absorbance of an EFS or an eclipsed ring pair differ from that of an isolated ring?
- Does the structure or fluorescence of an excimer in rigid media differ from that in fluid solution?
- Which rotational isomers of polychromophoric compounds generate intramolecular EFS?
- What is the proportion of intermolecular EFS to intramolecular EFS in bulk P2VN and other aryl vinyl polymers?

2.4.2 Evidence for the Sandwich Structure for Intermolecular Excimers

2.4.2.1 Theoretical Approaches

The theoretical approaches taken to calculate the binding energy of the excimer have been reviewed [68–70]. Most authors have assumed a sandwich structure for the excimer in which the ring planes are parallel and the molecular axes are aligned. By matching the calculated and experimental values of the excimer fluorescence peak, the interplanar distance of the excimer can be computed. All such calculations yield values of the interplanar distance which are 0.2–0.5 Å less than the ground-state van der Waals ring separation. For the naphthalene excimer, an interplanar distance of 3.3 ± 0.3 Å has been computed.

In order to properly account for repulsive forces in the excimer due to orbital overlap, the excimer state has been theoretically described [68–70] by configurational mixing of exciton-resonance states (^1L_a for naphthalene) and charge-resonance states. While the charge-resonance interaction is fairly isotropic and has a moderate inverse distance dependence, the exciton-resonance interaction is polarized and has a strong inverse distance dependence [69,71]. The latter interaction, which is predominant in all aromatic excimers except benzene and its derivatives, is responsible for maintaining the parallel arrangement of the molecular axes of naphthalene and other large arene compounds. Moreover, both interactions favor the sandwich structure of the excimer since this structure yields the smallest distance between molecular centers [69]. In calculations where certain conformations other than the symmetric sandwich structure were considered, the stability of the sandwich structure was confirmed [72] for the naphthalene excimer.

2.4.2.2 Volume and Extropy Change Upon Excimer Formation

The volume change ΔV associated with intermolecular excimer formation has been determined for naphthalene and various alkyl derivatives through the application of pressure [74]. For naphthalene and the two methylnaphthalene isomers, the value of $\Delta V = -16\ cm^3/mole$ was measured at room temperature. Assuming the sandwich structure for the excimer, and taking the projected area of the naphthalene molecule to be 60 Å^2, a decrease of 0.44 Å in the interplanar distance may be inferred,[74] in agreement with theoretical calculations of the electronic state of the excimer. For most of the dimethylnaphthalene isomers that were examined[74], ΔV was equal to -14.5 cm^3/mole. The reduced volume change for dimethylnaphthalenes relative to naphthalene is again consistent with the small increase in volume required by the methyl groups in the sandwich excimer of dimethylnaphthalene.

There is a substantial entropy decrease ΔS associated with intermolecular excimer formation, as given in the tabulation by Birks[71]. For all solvents (except 95% ethanol[75]), the value $\Delta S \approx -20$ cal/mole-K was observed for naphthalene and its derivatives. For comparison, the entropy of fusion of unsubstituted aromatic hydrocarbons such as naphthalene falls in the range of -8 to -15 e.u. The large loss of entropy in the intermolecular excimer formation process indicates a very constrained symmetric structure.

2.4.2.3 Fluorescence Considerations

Experimentally, the fluorescence spectra of concentrated solutions of monochromophoric compounds show a single excimer peak that is generally unaffected by alkyl substitution of the chromophores[71]. Moreover, it appears that mixed excimer can be formed between any two molecules, so long as they possess the same aromatic core[73]. Such mixed excimers (e.g. 1-methylpyrene and 4-methylpyrene) retain all the characteristics of the basic unsubstituted excimer. The only effect of alkyl substitution in pure and mixed excimers is a slight decrease in the binding energy, consistent with increased repulsion of opposing hydrogen and alkyl groups in the sandwich excimer structure.

The rate constant k_{TD} for fluorescence of the pyrene intermolecular solution excimer has been found to follow the relation $k_{FD} = n^2(k_{FD})_{n=1}$, where n is the the refractive index of the solvent[69]. The values of k_{TD} for the 1-methylnaphthalene excimer in ethanol at various temperatures are also consistent with the above relation[76]. The fact that $(k_{FD})_{n=1}$ is independent of solvent and temperature indicates that the excimer has a specific structure, according to Birks[69,71]. Experimentally, it was observed much earlier that $k_{FM} = n^2(k_{FM})_{n=1}$ for the polycyclic aromatic hydrocarbons, and that k_{FD}/k_{FM} is independent of solvent and temperature. Table 5 shows that agreement between independent investigators of the excimers of naphthalene compounds is not always good, as in the case of 1-methylnaphthalene.

Table 5. Radiative Rate Parameters for Naphthalene Compounds in Solution at Room Temperature (units of 10^6 s^{-1})

Compound	Solvent	k_{FD}	k_{FM}	k_{FD}/k_{FM}	Ref.
N[a]	95% EtOH[b]	0.9	2.3	0.39	[75]
1-methyl N	95% EtOH	1.1	2.9	0.38	[75]
	EtOH	2.6	2.0	1.33	[76]
2-methyl N	95% EtOH	1.1	3.4	0.32	[75]
1,6-dimethyl N	95% EtOH	1.3	3.6	0.36	[75]
	n-Heptane	1.4	5.0	0.28	[77]

[a] N = naphthalene; [b] EtOH = ethanol

In summary, all available evidence suggests that the intermolecular excimers of naphthalene compounds have a sandwich structure in which the ring planes are parallel and the molecular axes are aligned. While the intermolecular excimer appears to adopt the eclipsed sandwich structure in solution, there may be differences in the structure of excimers constrained by hydrocarbon links or by rigid matrices. These constrained excimers will be considered next.

2.4.3 Excimer Structure in Constrained Systems

Excimer fluorescence has been observed in a variety of systems in which intermolecular diffusion does not play a role in excimer formation. Five such systems involving the naphthyl chromophore will be discussed: (1) Crystals of aromatic hydrocarbons; (2) Glassy solutions or mixtures containing monochromophoric guest compounds; (3) Phane compounds, i.e. two chromophores held face-to-face by at least two hydrocarbon links; (4) Sandwich "dimers," which are chromophore pairs produced by photolysis of photodimers in rigid matrices; and (5) Bichromophoric compounds having a single saturated hydrocarbon linkage, which form intramolecular excimers as allowed by the rotational isomers of the linkage. In each case, we will utilize the intermolecular excimer formed in solution as the standard against which the properties of constrained excimers will be measured.

2.4.3.1 Crystals of Aromatic Hydrocarbons

Stevens [78] has reviewed the fluorescence properties of pure single crystals of aromatic compounds. In naphthalene, no excimer fluorescence is seen unless defects are introduced into the crystal. The fluorescence of defect-free crystals is most akin to that of naphthalene in dilute solution, although the crystal lattice does alter the fluorescence (and absorption) spectrum slightly. By contrast, only excimer fluorescence is emitted by pyrene crystals, at a band maximum within 500 cm^{-1} of that of the pyrene solution excimer [71]. Moreover, the fluorescence of crystalline pyrene is excimeric at all temperatures between 4 and 353 K. The absorption spectrum of the crystal, while slightly red-shifted relative to that of molecular pyrene, is still structured [78]. This suggests that the absorbing species is effectively monomeric. Kawakubo [86] has examined the fluorescence of some methyl- and dimethylnaphthalene solids. No excimer emission was detected in the room-temperature spectra of 2-methylnaphthalene and 2,6-dimethylnaphthalene crystals, or in the 77 K spectrum of a slowly-frozen sample of 1-methylnaphthalene. However, rapidly-frozen samples of 1-methylnaphthalene exhibited considerable excimer fluorescence, as did all samples of frozen 1,6-dimethylnaphthalene.

The different fluorescence spectra of naphthalene and pyrene crystals can be explained by examining the crystal structure. The naphthalene crystal [79] (Stevens's type A [78]) is monoclinic, composed of planes in which the long axes of the molecules are oriented roughly perpendicular to the plane. Each molecule has four nearest neighbors in the plane at a distance of 5.1 Å, and the angle between the short axis of the central molecule and that of a nearest neighbor is 55°. The two angles between the line of centers and the short axes within such a pair are about 25° and 100°. The pyrene crystal [79] (Stevens's type B [78]) is qualitatively similar to naphthalene, except that every naphthalene molecule is replaced in the lattice by a pair of pyrene molecules. The pyrene rings have their planes and axes parallel at an interplanar distance of 3.53 Å [71]. The pyrene molecules do not completely eclipse each other as the short axes are displaced by one C—C bond length whereas the long axes are coincident [71].

Birks [68] has proposed that the only change between the unexcited and excited pyrene pair is a reduction in the interplanar distance from 3.53 to 3.37 Å, i.e. that the pyrene excimer is not a completely eclipsed sandwich pair either in solution or in the crystal. This proposal is consistent with the observed similarity of the excimer band position for the crystal and solution environment, and with the emission of excimer fluorescence from the crystal even at 4 K. For naphthalene, the greater separation and the nonparallel structure of nearest-neighbor pairs in the crystal apparently prohibits the formation of the sandwich excimer during the naphthalene singlet monomer lifetime. Thus, no excimer fluorescence is observed from defect-free naphthalene crystals.

Kawakubo's fluorescence results [86] for methyl- and dimethylnaphthalene solids can be similarly related to the crystal structure. Both 2-and 2,6-substituted naphthalenes retain the same close-packed layer structure as seen in naphthalene. The only effect of the methyl substitution is to increase the crystal dimension along the naphthalene long axis [87]. Less is known about the crystal structures of 1- and 1,6-substituted naphthalenes, except that the 1-substituent requires a different packing pattern than naphthalene and that 1- and 1,6-substituted naphthalenes have much lower melting points than the 2-substituted naphthalenes. The absence of sandwich pairs in 2- and 2,6-substituted naphthalene crystals certainly explains the lack of excimer fluorescence in the crystal spectra. Presumably, such pairs are also absent in crystalline 1-methylnaphthylene, but they seem to be present in glassy 1-methylnaphthalene and in 1,6-dimethylnaphthalene solid.

2.4.3.2 Monochromophoric Compounds in Glassy Matrices

Given the distance requirements for excimer formation in the crystal, it would seem necessary to prepare glassy mixtures containing at least 1 mole/l of the guest compound in order to produce a few EFS, assuming a randomly-mixed solution. Unfortunately, it is quite difficult to prepare such concentrated low-temperature solvent glasses due to solubility limits at low temperature. Ferguson [80] has observed excimer fluorescence from frozen cyclohexane solutions containing as low as 10^{-4} mole/l pyrene. Although there are other glass-forming solvents in which pyrene is more soluble, studies on solvent glasses with concentrations above 5×10^{-3} mole/l have not appeared. For naphthalene, Jones and Calloway [46] measured an unusually low monomer lifetime for frozen cyclohexane solutions at 10^{-3} mole/l concentration. This was attributed [46] to crystallite formation, since the monomer lifetime of the naphthalene crystal is similarly low when compared to molecular naphthalene. The solubility problem was circumvented by utilizing 3-methylpentane or other solvent glasses [46]; molecular dispersions of naphthalene were obtained even at 10^{-2} mole/l concentration [81]. Regardless of crystallite formation, no naphthalene excimer fluorescence was observed in these studies.

Studies at higher guest concentration have been performed for glassy polymer hosts at room temperature. Pyrene has been dispersed in PMMA [82] and PS [83] at concentrations between 0.1 and 1.3 mole/l by solution casting from benzene or toluene. In both PMMA and PS, measurable excimer fluorescence was observed at 0.1 mole/l pyrene concentration. This is surprising, since the average distance between randomly-placed guest molecules is about 25 Å. While some pyrene aggregation is implied in the polymer hosts, several features of the excimer fluorescence distinguish these systems from the pyrene crystal. First, the excimer fluorescence response curves reach a maximum after a ~20 ns delay, while this delay is <1 ns in the pyrene crystal [84]. Second, the UV absorption spectrum of 1.0 mole/l pyrene in PMMA [82] is not broadened or red-shifted at wavelengths longer than 350 nm, whereas the spectrum of the pyrene crystal has these features [78]. Third, the monomer fluorescence decay curves of pyrene in PS [83] are neither mono- nor biexponential, but become increasingly non-exponential and show a shorter e^{-1} lifetime as the pyrene concentration is increased. Finally, the limiting lifetimes of excimer fluorescence at high pyrene concentration are 63–69 ns in PMMA [48,82] and 45–60 ns in PS [48,83], whereas the excimer lifetime of crystalline pyrene is 113 ns [84]. Note that the lifetime of the pyrene solution excimer is 65 ns in cyclohexane at room temperature [71].

These observations for pyrene-doped polymers have been interpreted as follows. The pyrene aggregates that allow excimer formation at 0.1 mole/l and higher concentrations are pairs of molecules, perhaps precursors to the crystal, which are formed from the random mixture by a weak ground-state interaction [82,83]. Such pairs form excimers which have the same lifetime and emission peak as solution excimers [82,83]. While Avis and Porter [82] have suggested that the delay in the excimer fluorescence response curves is due to the step of converting an excited pyrene pair to the excimer, Johnson's [83] interpretation of the delay seems more consistent with the non-exponential nature of the pyrene monomer fluorescence decay. He suggests that the delay in the excimer response is due to the step of transferring electronic excitation from a non-paired molecule to a pyrene pair [83]. The excited pair presumably forms the excimer within 1 ns, similar to the pyrene crystal. This interpretation is supported by the fluorescence spectra of pyrene/PMMA systems at 77 K, which do not differ substantially from those taken at room temperature [82]. A much more detailed analysis of Johnson's data that emphasizes the role of electronic energy transport has recently been published. [213]

Farid and co-workers [88] have investigated the effect of a glassy polymer host on the spectral position of the excimer emission peak produced by high concentrations of the compound methyl 4-(1-pyrenyl)-butyrate. The excimer peak position in a glassy polymer host was compared to the peak position in fluid solution for the following polymer hosts (and solvents): PS(toluene), PMMA(methyl isobutyrate), and poly-(vinyl benzoate) (methyl benzoate). The excimer emission peak of the pyrene compound in all three solvents occurred at about 20,800 cm^{-1}, but the emission peak in all three polymer hosts was blue-shifted about 1900 cm^{-1} relative to the solution value. This is in contrast to the behavior of unsubstituted pyrene in PMMA [82] and PS [83], whose excimer peak does not shift from the solution value.

Farid [88] did not report on the excimer lifetime of the pyrene compound in the systems that were studied. Nevertheless, they proposed that the blue shift of the excimer emission peak in glassy polymers relative to solution was due to improper orientation of the excimer components in the polymer matrix [88]. This proposal is supported by the observation [88] that the blue shift of the excimer peak for the pyrene compound in a low molecular-weight PS host was only 700 cm^{-1}, less than that observed for the other PS and PMMA hosts. In other work, [2] however, the 2-naphthyl excimer emission peak was found to be unaffected by the molecular weight of the PS or PMMA host.

Unfortunately, no studies have been made for naphthalene at high concentration in room-temperature polymer glasses. Siegel and Stewart [85] have prepared 0.2 mole/l samples of perdeuteronaphthalene in PS and PMMA by polymerizing mixtures of the host monomer and the guest. No excimer fluorescence was reported for these systems [85]. It seems reasonable to suppose that any naphthalene aggregates formed in concentrated systems would preferentially adopt the packing arrangement found in the crystal, i.e. the short axes of neighboring molecules will make an angle of 55° with each other while the long axes will be parallel [71]. The stability of a naphthalene sandwich pair is less certain. Further fluorescence studies including lifetime measurements may help to clarify the process of naphthalene aggregation in concentrated systems.

2.4.3.3 Phane Compounds

A unique set of materials for studying the effects of various orientations of a chromophore pair on the fluorescence of the pair are the phane compounds. For example, [3.2] (1,4) benzenophane represents two benzene rings held face-to-face by a $(CH_2)_2$ bridge attached at the 1-position of each ring, and a $(CH_2)_3$ bridge attached at the 4-position of each ring. A common name for [m.n] (1,4) benzenophane is [m.n] paracyclophane. The corresponding compound containing naphthalene rings instead of the benzene is of interest because two different forms are possible. The *anti*-isomer contains rings which are only 50% eclipsed, whereas the *syn*-isomer contains fully eclipsed rings. The photophysical properties and the structures of a wide variety of paracyclophanes are well-known, and many general aspects of bridged aromatic compounds are demonstrated by this series. For completeness we will briefly discuss the paracyclophanes, following the fine review given by Klöpffer [37]. However, since the properties of the napthalenophanes are more relevant to the systems emphasized in this work, they will be more extensively reviewed.

2.4.3.3.1 Paracyclophanes

A broad, structureless fluorescence emission is observed for [2.2], [3.3], and [4.4] paracyclophane, but only structured monomer emission is seen in [4.5] and [6.6] paracyclophane. The fluorescence properties of the [2.3], [2.4], [3.4], [3.6], [4.6], [5.5], and [5.6] paracyclophanes have not been reported, although the latter three would be expected to yield only monomer emission. The UV absorption spectra of all of the above paracyclophanes have been reported, and all [m.n] phanes for which both m and n are $\geqq 4$ have absorption spectra that are identical to 1,4-*bis* (4'-ethylphenyl)butane, the open-chain analog. The UV absorption spectra of other paracyclophanes become increasingly red-shifted and broadened in the order [3.6], [3.4], [2.4], [3.3], [2.3], and [2.2] paracyclophane.

The structure of the paracyclophanes explains many of the alterations that occur in the absorption spectra. In [2.2] paracyclophane, the inter-ring separation varies between 2.8 and 3.09 Å, and the benzene rings are bent out of the plane by 13°. This forced overlap lowers the first excited singlet state level by about 5800 cm^{-1}, as shown by the extended UV absorbance to the red of 320 nm. The peak of the broad fluorescence emission of [2.2] paracyclophane is also red-shifted 3300 cm^{-1} relative to the peak of the solution excimer of toluene at 31,300 cm^{-1}.

Distortion is less in the [3.3] paracyclophane, where the inter-ring separation varies between 3.1 and 3.3 Å, and the benzene rings are only bent by 6°. In the crystal, the hydrocarbon links adopt the chair conformation, and the benzene rings are displaced by about 0.5 Å from the sandwich structure, although the rings remain parallel. Some ground-state overlap occurs between the rings, since the UV absorbance extends to the red of 305 nm. The peak of the broad fluorescence emission of [3.3] paracyclophane appears at the same position as that of [2.2] paracyclophane.

The fluorescence bands in [2.2] and [3.3] paracyclophanes should not be thought of as "true" excimer fluorescence since the ground state in these phanes is not free from interaction. In fact, the low-temperature fluorescence spectrum of [2.2] paracyclophane has been reported to show considerable structure, although this was not observed for [2.2] or [3.3] paracyclophanes at low temperature in a more recent report [89a].

The inter-ring separation in [4.4] paracyclophane has been calculated to be 3.73 Å, assuming normal bond angles and planar benzene rings. At this distance, there is no ground-state overlap, and the UV absorbance does not extend past 280 nm. Nevertheless, the peak of the excimer fluorescence intensity of [4.4] paracyclophane is red-shifted 1900 cm^{-1} relative to the peak of the solution excimer of toluene at 31,300 cm^{-1}. Neither the excimer lifetime nor the excimer fluorescence response function have been reported for any of the excimer-forming paracyclophanes, so little is known about the kinetics of excimer formation in these compounds.

2.4.3.3.2 Naphthalenophanes

The naphthalenophanes that have been synthesized to date are listed in Table 6, in order of their discovery. The [m.n] isomers for which m,n > 3 have not yet been synthesized. References for the UV absorbance, fluorescence, and other properties of existing naphthalenophanes are given in Table 6. The UV absorption spectra of all the naphthalenophanes are red-shifted and broadened relative to their respective open-chain analogs, similar to the [2.2] and [3.3] paracyclophanes. Moreover, broad and structureless emissions have been observed for the naphthalenophanes in all references cited in Table 6 except one. [107] The structural aspects of naphthalenophane photobehavior will be discussed in detail in the following paragraphs.

The naphthalenophanes that are fully eclipsed, i.e. the *syn*-[2.2](1,4), achiral [2.2](1,5), achiral [3.3](2,6), *syn*-[3.3](1,4), and *syn*-[3.2](1,4) isomers, share certain traits in absorption and fluorescence. The UV absorbance spectra of these compounds between 260 and 310 nm retain all of the structure shown in the spectra of the open-chain analogs. Also, new absorption shoulders not seen in the open-chain spectra appear strongly at 245 and weakly at 340 nm. The fluorescence peak of these fully eclipsed naphthalenophanes occurs near 22,000 cm^{-1}, as seen in Table 7. This represents a red shift of 2600 cm^{-1} relative to the solution excimer of the dimethylnaphthalenes. [71]

The remaining naphthalenophanes in Table 6, which are noneclipsed, show little vibrational structure in their UV absorption spectra relative to the open-chain analogs.

Table 6. Naphthalenophane Bibliography

Compound[a]	Chem. Abstr. Registry No.	UV Abs. References	Fluorescence References	Other References
[2.2](2,7)N	—	—	—	89b)
anti-[2.2](1,4)N	14724-91-5	90–92, 106)	70, 92, 93)	94)
syn-[2.2](1,4)N	23284-44-8	91, 92, 106)	92)	94)
chiral[2.2](2,6)N	67374-99-6	95, 96)	70)	—
achiral[2.2](2,6)N	—	—	—	97)
chiral[2.2](1,5)N	54835-57-3	98b, 100)	99b, 100)	—
achiral[2.2](1,5)N	54911-02-3			
[2.2](2,6)(2′,7′)N	59456-89-2	—	—	101, 102)
chiral[3.3](2,6)N	67500-25-8	103, 104)	—	—
achiral[3.3](2,6)N	67529-23-1			
anti-[3.3](1,4)N	70672-12-7	105, 106)	106, 107)	—
syn-[3.3](1,4)N	70748-69-5			
anti-[3.2](1,4)N	78020-44-7	108)	—	—
syn-[3.2](1,4)N	78007-69-9			
‖-[3.3](1,5)(2′,6′)N	78007-62-2	104)	—	—
⊥-[3.3](1,5)(2′,6′)N	78007-63-3			

[a] N = naphthalenophane;
[b] The assignments of the diastereoisomers are reversed. See Ref. 100)

New absorption shoulders appear strongly at 250 and 345 nm. Given this evidence of ground-state interaction, the fluorescence band of the noneclipsed naphthalenophanes should be red-shifted below the peak emission of the dimethylnaphthalene solution excimer. In fact, the emission of chiral [2.2](2,6) naphthalenophane is blue-shifted 900 cm^{-1}, and the emissions of the *anti*-[2.2](1,4), *anti*-[3.3](1,4), and chiral [2.2](1,5)

Table 7. Naphthalenophane Fluorescence

Compound[a]	Solvent[b]	T, K	ν_D, cm^{-1}	Ref.
anti-[2.2](1,4)N	EPA	?	24,800	92)
	PMMA	4.2	23,800	70)
	CH	298	24,000	93)
	A	298	24,300	93)
	EA	77	23,800	93)
syn-[2.2](1,4)N	EPA	?	21,700	92)
chiral[2.2](2,6)N	PMMA	4.2	25,500	70)
chiral[2.2](1,5)N	PMMA	1.3	23,000	99, 100)
achiral[2.2](1,5)N	PMMA	1.3	22,200	99, 100)
anti-[3.3](1,4)N	EPA	?	23,800	106)
	CH	298	22,900–24,400	107c)
syn-[3.3](1,4)N	EPA	?	23,200	106)
	CH	298	21,500	107)

[a] N = naphthalenophane;
[b] EPA = ether, isopentane and ethanol 5:5:2 v/v/v; EA = ether and ethanol 1:2 v/v; CH = cyclohexane; A = acetonitrile;
[c] The spectrum shows structure at 350 nm, probably due to photoreaction products

isomers are red-shifted 600, 600, and 1600 cm^{-1}, respectively. The binding energy of these noneclipsed ring pairs is evidently less than in the eclipsed isomers.

The (1,4) substituted naphthalenophanes undergo [4 + 4] photocycloaddition when irradiated at $\lambda > 280$ nm, in addition to fluorescence. This photoreaction is competitive with fluorescence, and requires a conformational change that can be suppressed at low temperature [93]. The few reports of the lifetime or quantum yield of naphthalenophane fluorescence indicate the effects of photocycloaddition. For the *anti*-[2.2](1,4) isomer, $k_{FD}/k_D = 0.021$ in cyclohexane [93]; the lifetime of *syn*-[3.3](1,4) naphthalenophane fluorescence was given as 15.3 ns [107]. Both values are low relative to the naphthalene solution excimer ($k_{FD}/k_D \approx 0.2$; $\tau_D \approx 80$ ns [71]), and this may be due in part to the photoreaction of the (1,4) naphthalenophanes.

In conclusion, there is much to be done in characterizing the photophysics of naphthalenophanes. The fact that the eclipsed isomers emit at lower energies relative to the noneclipsed isomers is in accord with the assignment of the parallel-plane sandwich structure to the naphthalene solution excimer.

2.4.3.4 Photodimers

There are a wide variety of aromatic compounds that form photodimers. Because these dimers can be photolyzed back to the starting materials by irradiation with high-energy UV light ($\lambda < 250$ nm), this property permits pairs of molecules, or sandwich dimers, to be formed and studied in rigid matrices. Chandross [109] introduced this technique in the study of sandwich dimers formed from the photolysis of dianthracene in rigid solvent glasses. The numerous studies of sandwich dimers of anthracene and its derivatives have been reviewed by Birks [69, 71]. In one of the many studies contributed by Ferguson and co-workers [110], the fluorescence spectrum and lifetime of anthracene sandwich dimers were followed as the solvent glass was slowly warmed.

Intermolecular photodimerization of alkylnaphthalenes such as 1-methylnaphthalene [10] does not occur, even in the pure liquid. However, the intramolecular process has been observed in bis(1-naphthyl)alkanes and ethers in which the link between the rings contains three atoms. Chandross and Dempster [10] were the first to observe [4 + 4]photocycloaddition occurring in 1,3-bis(1-naphthyl) propane (ααDNP) to yield the *syn*-photodimer exclusively. The same authors [111] studied the sandwich dimer obtained by photolysis of the photodimer in a methylcyclohexane glass at 77 K. The fluorescence of the ααDNP sandwich dimer was identical to the excimer emission of ααDNP in fluid solution, and the UV absorbance of the sandwich dimer was consistent with a sandwich pair configuration [111]. The same sample, when thawed and refrozen, showed the normal UV absorbance of ααDNP. Because photodimerization, like excimer fluorescence, did not occur for 1,2-bis(1-naphthyl) ethane or 1,4-bis(1-naphthyl)butane, it was suggested that excimer formation was a necessary precursor to photodimerization in these compounds [10]. No photoreaction was observed [10] for 1,3-bis(2-naphthyl)propane, even though this compound readily forms excimers. Evidently, photodimerization requires additional electronic factors beyond those involved in excimer formation.

Intramolecular photodimerization in P1VN and ααDNP was studied for cyclohexane, dichloromethane, and benzene solvents [117]. The conversion to photodimer was measured from the decline in sample absorbance at 295 nm. The maximum P1VN conversions obtained were 70, 20, and 40% with respect to the solvents given above. The low conversions in benzene and dichloromethane were not fully explained. Moreover, the creation and study of the sandwich dimer in P1VN by the Chandross technique [111] was not pursued. The possibility of controlling the number of sandwich dimers in rigid solutions of P1VN is most intriguing.

While the photodimerization of bis(1-naphthylmethyl) ether was acknowledged somewhat earlier [39], the photodimers were first characterized and the quantum yield of the dimerization determined by Todesco et al. [112]. Both the *syn*- and *anti*-photodimers were formed in roughly equal amounts, and the quantum yield for formation of the *anti*-dimer was independent of solvent. However, the quantum yields for formation of the *syn*-dimer and for excimer fluorescence were found to vary with solvent such that their sum was independent of solvent. The fact that irradiation of 1,3-bis(1-naphthyl)-1-propanol yields only the *syn*-photodimer [113] indicates that the conformational properties of oxygen are largely responsible for *anti*-dimerization in the ether compound. The possibility of photodimerization was unfortunately not considered in the fluorescence studies of protonated bis (1-naphthylmethyl) amine [115], 1,3-bis(4-methoxy-1-naphthyl) propane [116], and *meso*-bis(1-(1-naphthyl)-ethyl) ether [13].

In a later paper, Todesco et al. [114] separated the *syn*- and *anti*photodimers of bis(1-naphthylmethyl) ether, and studied the sandwich dimers obtained by photolysis in an ethanol solid solution at 77 K. The same fluorescence spectrum, with a broad band at 415 nm, was observed for both sandwich dimers. This is unusual, because the emission maxima of *syn*- and *anti*-(1,4) naphthalenophanes differ by about 2,000 cm^{-1} (see Table 7). Structures of *syn*- and *anti*-excimers were proposed [114] in which the parallel naphthyl rings are shifted so that the aromatic hydrogens along the long axis are staggered. Thus, by reducing ring overlap in the *syn*-excimer and increasing the overlap in the *anti*-excimer, the same interaction can be achieved by both forms. This proposal was consistent with the monomer and excimer fluorescence decay behavior of bis(1-naphthylmethyl)ether in solution. [114] However, the UV absorption spectrum and fluorescence lifetime of the sandwich dimers of this ether compound need to be determined to confirm the two-excimer hypothesis.

It is clear that the sandwich-dimer studies discussed above apply to P1VN, not P2VN, since no photodimerization has been observed in bis(2-naphthyl)alkanes [10] and ethers [39]. Nevertheless, the UV absorbance of naphthyl sandwich dimers, like that expected for [3.4] or [3.5] naphthalenophanes, differs from that of isolated molecules only for $\lambda > 325$ nm. The same slight difference in UV absorbance probably occurs for excimer-forming sites.

2.4.3.5 Bichromophoric Compounds With Single Flexible Three Atom Linkages

2.4.3.5.1 Survey of Interunit Linkages

The observation that only the 1,3-diarylpropanes show significant amounts of excimer fluorescence, known as Hirayama's "n = 3 rule", originates from the fact that the parallel, fully-eclipsed sandwich structure formed between opposing phenyl, 1-naphthyl and 2-naphthyl groups is most stable and that a linkage with three carbon atoms can provide this excimer structure without incurring bond-angle distortions or unstable rotational isomers. This rule was refined after a comprehensive study [119] of α,ω-bis(1-pyrenyl) alkanes, covering all linkages from ethane to hexadecane. While the ratio I_D/I_M was largest for n = 3, the only compounds that did *not* show excimer fluorescence were n = 2, 7, and 8. In fact, I_D/I_M for the n = 4 and 13 compounds was reduced only 50% relative to the n = 3 compound. Pyrene excimer fluorescence has been recently observed [120] for polystyrene molecules having n = 60–2,000 and which are end-capped with pyrene.

The reason that the pyrene compounds exhibit excimer fluorescence over a broader range of linkages than the phenyl or naphthyl compounds is that the monomer lifetime of pyrene is over six times longer, and the pyrene excimer binding energy is ~40% larger [71]. The longest linkage in an α,ω-bis(2-naphthyl)alkane to be examined [40] was n = 12. This compound did not show excimer fluorescence between 173–298 K, and it seems unlikely that excimers will occur in longer-linkage 2-naphthyl compounds.

Hirayama's rule can be extended to compounds in which the chain connecting the chromophores contains one or more heteroatoms. Excimer fluorescence has been observed in numerous compounds having the general structure R(C—X—C)R. For X = 0, many different chromophores R have been studied: phenyl [115,123,124], 1-naphthyl [39,112,114,115], and 2-naphthyl [39]. Also, the diastereometric compounds bis (1-arylethyl)ether containing the 1-naphthyl or 2-naphthyl chromophore have been studied [13]. A number of diphenyl substances having nitrogen in the linkage have been prepared: X = N—CHO [123], N—$COCH_3$ [115], and NH_2^+ [115,123]. The bis(1-naphthyl) compound having X = NH_2^+ has also been prepared [115,125]. Finally, aryloxy-substituted cyclophosphazenes, which possess three $_{R-O}>P<_{O-R}$ groups, have been synthesized and characterized [131] for R = phenyl, 1-naphthyl, and 2-naphthyl.

All the above compounds yield excimer fluorescence when excited in room-temperature solution. However, because the rotational potential of the C—X bond and the nonbonded interactions of the substituents of the X atom differ from those of the C—C bond [126], the amount of excimer fluorescence from R(C—X—C)R differs from that of R(C—C—C)R. The heteroatom X can also influence the rotational state of the side groups R, as illustrated by the formation of the anti-photodimer in bis(1-naphthylmethyl)ether [112], but not in 1,3-bis(1-naphthyl)propane [10]. Finally, compounds having n ≠ 3 may exhibit excimer fluorescence, if the linkage contains one or more heteroatoms. For example, the—C—O—C—C— linkage in α,ω-bis(2-naphthyl) compound allows excimer fluorescence to be observed in room-temperature solution [39].

2.4.3.5.2 Thermodynamics

The parameters ΔH, ΔS and ΔV associated with intramolecular excimer formation in bichromophoric compounds can be determined by applying a modified version [37] of the well-known kinetic scheme for intermolecular excimers [71]. Such a study of 1,3-bis(2-naphthyl)propane in methylcyclohexane gave ΔH = −5.7 kcal/mole and ΔS = −9.9 cal/mole-K [121]. Relative to the intermolecular excimer of naphthalene and its derivatives [71], the entropy decrease for the intramolecular excimer is half as large. The enthalpy change, on the other hand, is only slightly less exothermic than that of the intermolecular excimer. As expected, the propane linkage between the naphthyl ring increases the stability of the excimer at high temperature. The only report on ΔV for an intramolecular excimer concerned the compound 1,3-bis(*N*-carbazolyl)propane, for which $\Delta V = -6.2\ cm^3/mole$ in ethanol [122]. Thus volume decrease is smaller than the $-16\ cm^3/mole$ value seen for the naphthyl intermolecular excimer, again reflecting the presence of the linkage between the rings.

The position of the excimer fluorescence peak of the 1,3-dinaphthylpropanes depends on the naphthyl ring substitution. For the 2-naphthyl isomer [9,127], the excimer peak energy is $\nu_D = 24{,}800 \pm 200\ cm^{-1}$, which is the same as that of the intermolecular excimer of 2-substituted naphthalenes [71]. However, the excimer peak of the 1-naphthyl isomer [9,107,115,129] is red-shifted about $1{,}000\ cm^{-1}$ relative to the corresponding intermolecular excimer [71], for which $\nu_D = 25{,}000\ cm^{-1}$. The red shift is probably due to increased repulsion between the naphthyl rings in the excimer-forming site of the 1-naphthyl isomer, because bond-angle deformations in the linkage do not relieve ring overlap as effectively as in the' 2-isomer. Measurement of ΔH for the 1,3-bis(1-naphthyl)propane excimer is needed to fully explain these differences, however.

2.4.3.5.3 Existence of Multiple Excimers

Only one excimer fluorescence peak and lifetime were observed [12] for R—CH_2—CH_2—CHX-R, where R = 2-naphthyl and X = H [121], methyl, dodecyl, or β-phenylethyl. The same was also true for *meso*- and *dl*-2,4-bis(2-naphthyl)pentane [12]. A recent report [128] attributing the fluorescence peaks at 345, 363, and 381 nm in the spectrum of poly(2-*tert*butyl-6-vinylnaphthalene) to a second excimer species must therefore be considered invalid. The polymer undoubtedly contains an in-chain impurity with vinylnaphthalene conjugation, as discussed in Section 2.1.9.2 for a similar claim involving P2VN [33].

There is evidence both for and against the contention that only one excimer fluorescence peak and lifetime is possible for bis(1-naphthyl) compounds having n = 3. Studies of 1,3-bis(1-naphthyl)propane [9,107,115,129] and bis(1-naphthylmethyl)ether [39,114,115] in various solvents over a range of temperatures have found only one excimer fluorescence peak and decay rate (even though there appear to be two possible excimer structures in the ether compound [114]. On the other hand, fluorescence peaks attributed to two excimer types have been recorded at 28,200 and 26,700 cm^{-1} for *meso*-bis(1-(1-naphthyl)-ethyl)ether [13], and at 27,000 and 24,400 cm^{-1} both for the compound 1,3-bis(4-methoxy-1-naphthyl)propane and for 1,3-bis(4-hydroxy-1-naphthyl)propane [116].

Second, although the compounds 1,3-bis(1-naphthyl)propane and the bis(2-naphthyl)isomer both showed a single excimer fluorescence band in toluene at 303 K, the excimer bandwith at half height of the 1-naphthyl isomer was 600 cm^{-1} larger than the value of 3790 cm^{-1} for the 2-naphthyl isomer [127]. This raises the possibility that the emission of the 1-naphthyl isomer contains a weak, unresolved contribution from a second excimer. Finally, in a study of the fluorescence lifetimes of P1VN in dichloromethane [130], two excimer emissions at 25,300 and 23,150 cm^{-1} were detected, having τ_D = 14 and 22 ns, respectively. The net excimer fluorescence peak appeared at 24,000 cm^{-1}, however [130]. A re-examination of the bis(1-naphthyl) compounds by time-resolved fluorescence spectroscopy [1b] is needed to resolve the subtleties of the excimer or excimers in these systems.

2.4.3.5.4 Conformational Statistics

The fluorescence behavior of the bis(2-naphthyl) compounds implies that the predominant intramolecular excimers in P2VN are formed between nearest-neighbor rings ($n = 3$; $n \neq 5, 7, 9, 11$ and 13). Because the fluorescence peaks of the inter- and intramolecular 2-naphthyl excimers are identical, the latter presumably has the same fully-eclipsed sandwich structure as the former. The question of the conformation of 2-naphthyl the long axis of the excimer relative to the plane containing the three atoms of the linkage will be discussed later. More important is an enumeration of the rotational conformations of the dyads of an atactic P2VN chain which give the intramolecular excimer.

The conformational statistics of asymmetric vinyl chains such as P2VN are well-known [126]. The rotational conformers of isotactic (*meso*) dyads are entirely different from those of syndiotactic (*dl*) dyads. Frank and Harrah [132] have described each of the six distinct conformers for *meso* and *dl* dyads, using the t, g^+ and g^- nomenclature of Flory [126]. Excimer-forming sites (EFS) are found in the tt and g^-g^+ *meso* states, and in the degenerate tg^-, g^-t *dl* state. Because the rotational conformers of compounds such as 1,3-bis(2-naphthyl)propane do not match those of either the iso- or syndiotactic dyads of P2VN, the propane compounds make poor models of aryl vinyl polymers. However, the rate constants of fluorescence and decay of the intramolecular excimer in polymers can usually be determined from the propane compounds (but see the exceptional case of PVK and its models [133]).

The *meso* and *dl* isomers of 2,4-diarylpentanes are capable of modelling all aspects of the behavior of aryl vinyl polymers, except for the interaction between adjoining dyads and other long-range interactions. Model pentanes containing the phenyl [6,124,134,135], 2-naphthyl [12], or *N*-carbazolyl [133] chromophores have been prepared and studied in solution by fluorescence spectroscopy. For the phenyl and 2-naphthyl compounds, the ratio I_D/I_M at room temperature of the *meso* isomer is about 5–10 times as large as that for the *dl* isomer. The carbazole compounds are unique in that the *meso* isomer shows low-energy ($\lambda_D = 420$ nm) excimer fluorescence, and the *dl* isomer shows high-energy ($\lambda_D = 370$ nm) excimer fluorescence. These observations can be related to the structure of the predominant rotational conformers of *meso*- and *dl*-2,4-diarylpentanes.

Conformational free energy calculations based on NMR measurements from the diastereomers of 2,4-diphenylpentane have shown [136] that for the *meso* isomer, the degenerate tg^-, g^+t state is most stable, the tt state (EFS) is 1.8 kcal/mole higher, and all other states are sufficiently high as to be neglected. For the *dl* isomer, the tt state is most stable, the g^+g^+ (and perhaps the degenerate tg^+, g^+t) state is <1 kcal/mole higher, and all other states can be neglected. The latter includes the tg^-, g^-t EFS state, which is >3 kcal/mole above the tt state. Thus, the $tg^- \rightarrow tt$ barrier to *meso* excimer formation is much less than the $tt \rightarrow tg^-$ barrier to *dl* excimer formation, and so $(I_D/I_M)_{meso} \gg (I_D/I_M)_{dl}$.

Recent conformational calculations made on the 2,4-bis(2-naphthyl)-pentanes [137] gave results similar to the diphenylpentanes. The tt state (EFS) in the 2-naphthyl *meso* isomer was found to be 2.3 kcal/mole higher than the tg^-, g^+t state, consistent with a larger repulsion between naphthyl rings than for phenyl rings. It is apparent from molecular models, and has been confirmed by Seki et al. [138,139], that the

2-naphthyl rotational state in which the naphthyl C(2)–C(3) bond is eclipsed with the backbone αC-H bond has the same conformational energy (within 0.3 kcal/mole) as the rotational state 180° apart. The tt *meso* state in which the naphthyl rings have opposite rotational states is probably more stable than the one having the same rotational states for both naphthyl rings, but no energy calculations have been made. Despite the minor complications induced by the rotational state of the 2-naphthyl group, the *meso*-2,4-bis(2-naphthyl)pentane shows a much larger value of the fluorescence ratio than the *dl* isomer.

Finally, NMR measurements on the 2,4-bis(N-carbazolyl) pentanes [133] indicate that the same rotational states as seen in the phenyl and 2-naphthyl compounds are applicable. The relative stability of the tt *meso* state correlates with the low-energy "sandwich" excimer fluorescence emitted at 420 nm from the *meso* isomer. Because of the size and shape of the *N*-carbazolyl group, the phenyl portions of the carbazoles in the tt *dl* isomer overlap. The high-energy excimer ($\lambda_D = 370$ nm) evidently corresponds to this structure in the *dl* isomer. Conversely, the high-energy excimer cannot be formed in the *meso* isomer because the tg^+, g^-t conformer is energetically disfavored. Also, the low-energy excimer in the *dl* isomer requires the unfavorable tg^-, g^-t conformer. This explains why low- and high-energy excimer fluorescence is seen exclusively from the *meso* and *dl* isomers, respectively.

While the study of diarylpentanes is helpful in understanding the conformational behavior of aryl vinyl polymers, a simple weighting of the properties of the model compounds by the tacticity of the polymer does not yield the properties of the polymer. For example, the presence of *dl* dyads surrounding a *meso* dyad will suppress the tt conformer in the *meso* dyad [140]. Thus, in order to obtain the fraction of tt *meso* conformers within an atactic P2VN sample, it is necessary to resort to a Monte Carlo calculation utilizing an extended product of statistical weight matrices [126].

The statistical weights for the P2VN chain are generally similar to those of PS. However, the tt *meso* state for the P2VN dyad was found to be ~0.5 kcal/mole less stable than for the PS dyad, as noted earlier. Refinement of the statistical weights for P2VN would require data on the end-to-end distance of atactic and isotactic P2VN samples in theta solvents, and on stereochemical equilibria and conformer populations in P2VN oligomers. Because these data are generally unavailable and the tacticity of the P2VN samples in this work is unknown, we will take the calculated fraction of tt *meso* dyads in an atactic (45–50% *meso*) PS sample as an upper limit on the tt *meso* fraction in P2VN.

The tt *meso* dyad fraction for an atactic PS sample at 298 K was computed [140, 141] to be 0.026, based on the statistical weights of Yoon et al. [142]. If it is assumed that the two rotational states of the 2-naphthyl ring are isoenergetic, then there is a 50% probability that the naphthyl groups in a tt *meso* dyad will be fully eclipsed. Thus, the adjacent intramolecular EFS dyad fraction, q_D, for an atactic P2VN sample is estimated to be 0.013 at 298 K. The rotational barrier for the 2-naphthyl ring in a tt *meso* dyad is assumed to be sufficiently large to prevent conversion of a non-eclipsed ring pair to the EFS in rigid solution.

2.4.3.6 Polychromophoric Compounds Without Three-Atom Linkages

The question remains whether intramolecular excimer formation is possible between two aromatic rings separated by $n > 13$ carbon atoms. When n is large, the problem becomes similar to the problem of determining how much intermolecular excimer formation is possible in systems containing >1% of an aryl vinyl polymer. The fluorescence properties of polychromophoric compounds in which the n = 3 ring spacing is absent are useful in answering these questions. Such compounds include head-to-head polymers made from an aryl vinyl monomer, alternating copolymers, and various homopolymers having pendent rings which are separated from the backbone by one or more atoms.

Head-to-head addition of an aryl vinyl monomer produces a polymer in which the ring spacings of n = 3, 7, 11, ... are absent. However, the value of n = 13, which was favorable to intramolecular excimer formation [119] in α,ω-bis(1-pyrenyl) alkanes, is present in head-to-head polymers, so that the potential for nonadjacent excimer formation is maximized.

The only head-to-head polymer which has been examined for excimer fluorescence is polystyrene [25]. Unfortunately, the synthetic route to this polymer leaves a number of stilbene-based structures in the sample, which have a lower-energy singlet state than either PS monomer (285 nm) or excimer (330 nm). Thus, fluorescence from these intrinsic stilbene traps was seen in the spectra of head-to-head PS in pure films and, to a lesser extent, in fluid solution. In the latter, the fluorescence of PS monomer was predominant, and the small amount of stilbene fluorescence was increased when a nonsolvent (methanol or cyclohexane) was added to the 2-methyltetrahydrofuran solution. In films of the polymer, stilbene fluorescence was the major spectral band, although some PS excimer fluorescence was also present in the spectrum. No monomer fluorescence at 285 nm was detected from films. Given the impure nature of the head-to-head PS sample, no conclusions on excimer formation in these systems could be drawn.

In the second category, polymers with the repeat unit [CH_2—CRR′—$(CH_2)_m$ CRR′—CH_2], where R = phenyl, m = 3–6 or 10, and R′ = hydrogen or methyl, were synthesized by Richards et al. [143]. The fluorescence of the R′ = H compounds [144] and the R′ = CH_3 compounds [25, 143] were studied in fluid and rigid solution and in pure films. Although no spectra were given for the R′ = H compounds, these were stated [144] to have no excimer emission at 330 nm in fluid solution nor in pure films. A similar report was made [25] for the R′ = CH_3 compounds in 2-methyltetrahydrofuran solution at room temperature, and in such solutions to which methanol had been added to the point of opalescence. These results were confirmed for the R′ = CH_3 compounds in solution, and the spectra of pure films did not show significant amounts of excimer fluorescence at 330 nm [145]. However, an extraneous emission at 310 nm in the film spectra made quantitative measurement of the 330 nm excimer band impossible.

Despite the technical problems in the latter film study, we conclude that there is no intramolecular excimer formation in the compounds of Richards et al. [143], and probably little intermolecular excimer formation in the pure films. The absence of an effect of solvent power [25] on the possible excimer fluorescence of the R′ = CH_3 polymer may not be significant, since little change in the coil dimensions would be expected for the short (~300 backbone atoms) polymers [143] which were studied. Additional work is needed on the fluorescence of such polymers having higher molecular weights, different aryl substituents (R = 2-naphthyl, for example), and fewer adventitious impurities.

A variety of alternating copolymers, made from the monomers styrene [55], 2-vinylnaphthalene [55], or *N*-vinylcarbazole [146] and a nonfluorescent monomer such as methyl methacrylate or fumaronitrile, have been characterized by fluorescence spectroscopy. In dilute dichloromethane solution, the polymers P(S-*alt*-MMA) and P(2VN-*alt*-MMA) exhibited no signs of excimer fluorescence at 330 and 400 nm, respectively [55]. Moreover, the addition of cyclohexane to the above copolymer solutions did not alter the fluorescence spectra. The carbazole copolymers P(VK-*alt*-fumaronitrile), P(VK-*alt*-diethyl fumarate), and P(VK-*alt*-diethyl maleate) all gave fluorescence spectra similar to *N*-ethylcarbazole in dilution [146]. There was little trace of either low-energy or high-energy excimer fluorescence at 420 or 370 nm, respectively.

The pure-film spectra of the alternating copolymers have been reported only for P(S-*alt*-MMA) and P(2VN-*alt*-MMA) [55]. For the former polymer, a fluorescence band similar to toluene was observed, distinguished only by a slight broadening at $\lambda > 290$ nm. The P(2VN-*alt*-MMA) film spectrum was described as having a maximum at 365 nm, with appreciable fluorescence intensity at 425 nm. This emission, which could not be positively identified as excimer fluorescence, was attributed to an impurity by Fox et al. [55]. If the results for P(S-*alt*-MMA) are representative of all the copolymer film spectra, they indicate that very few intermolecular EFS are formed.

The third group of polychromophoric compounds to be discussed are homopolymers in which the pendant rings are separated from the backbone by one or more atoms. The polymers of allyl arenes, which lack only the n = 3 ring spacing of aryl vinyl polymers, have been studied very little. The fluorescence spectrum of poly(1-allylnaphthalene) in dilute dichloromethane solution has been reported [28]. Like 1-ethylnaphthalene, the maximum intensity was seen at 337 nm, but a weak, broad shoulder was also recorded for the polymer at 410 nm. The fluorescence ratio I_D/I_M for poly(1-allylnaphthalene) was only 1/100 *th* the value for P1VN [28]. The excimeric nature of the 410 nm emission in the allyl-based polymer has not been confirmed, since neither the lifetime nor the excitation spectrum of this fluorescence band are known.

A wide variety of arylalkyl methacrylate polymers, having the structure [CH_2—C(CH_3)(COOR)], have been the subject of fluorescence studies. We will consider only the phenyl- and naphthyl-containing polymers. In a study of the benzyl, 2-phenylethyl, or 3-phenylpropyl methacrylate polymers, excimer fluorescence was not seen in dichloromethane solution, nor in ethyl acetate solution [43]. Poly(phenyl methacrylate) was not studied, since this polymer undergoes a rapid photoreaction. Excimer fluorescence from immiscible, solvent-cast blends of poly(benzyl methacrylate) (PBzMA) and PMMA was characterized [147] by the ratio I_D/I_M, which increased with concentration in the same manner as was observed for immiscible blends of PS and poly(vinyl methyl ether) [148]. However, I_D/I_M for pure PBzMA was only 1/10 *th* the value for pure PS, and I_D/I_M approached zero as the concentration of PBzMA was reduced, in contrast to the nonzero limit observed in PS blends. It is clear that PBzMA and the other phenylalkyl methacrylate polymers are incapable of intramolecular excimer formation in both fluid and rigid systems.

Guillet and co-workers have made extensive studies of the 1-naphthylalkyl methacrylate polymers, including the 1-naphthyl [149,150,152], 1-naphthylmethyl [151], and 2-(1-naphthyl)ethyl [151] derivatives. Poly-(1-naphthyl acrylate) has also been investigated [150,152]. All of these polymers exhibit excimer fluorescence in solution. For comparable molecular weights in tetrahydrofuran solution, I_D/I_M declines in the order 1N acrylate ≈ (1N)methyl methacrylate > 1N methacrylate > (1N)-ethyl methacrylate.

The excimer interaction is probably limited to nonadjacent rings in the latter three polymers, based on the observations that bis(1-naphthylmethyl) glutarate and bis (2-(1-naphthyl)ethyl) glutarate do not show excimer fluorescence [151], and that I_D/I_M of the three polymers in nonquenching solvents decreases as the intrinsic viscosity (i.e. the coil size) increases [149-151,153]. In contrast, adjacent intramolecular excimer formation predominates in poly(1-naphthyl acrylate), since bis(1-naphthyl) glutarate exhibits excimer fluorescence [153]. Moreover, there is little solvent effect on I_D/I_M of poly(1-naphthyl acrylate) [150]. Additional data on more realistic bichromophoric compounds is needed to elucidate the nature of the intramolecular excimer in these polymers, however. The solvent effect on I_D/I_M may also be due solely to changes in the energy migration rate, and not to differences in the excimer formation rate. Finally, the occurrence of a photo-Fries reaction in the 1-naphthyl acrylate polymer and the 1-naphthyl methacrylate polymer should be noted [152].

The 2-naphthylalkyl methacrylate polymers, such as the 2-naphthyl [55,152], 2-naphthylmethyl [154], 1- [155] and 2-(2-naphthyl)ethyl [154], and 3-(2-naphthyl)propyl [154] derivatives, and poly(2-naphthyl acrylate) [152] have been studied. The behavior of the 2-naphthyl polymers is similar to the 1-naphthyl polymers in the following areas. First, excimer fluorescence is observed for all the 2-naphthyl polymers in solution, and I_D/I_M declines in the order 2N acrylate > 2N methacrylate ≈ (2N)methyl methacrylate > (2N) propyl methacrylate > 2-(2N)ethyl methacrylate > 1-(2N)ethyl methacrylate. Second, the value of I_D/I_M for any 2-naphthylalkyl methacrylate or acrylate derivative is roughly equal to that of the corresponding 1-naphthyl derivative, in room-temperature tetrahydrofuran solution. Third, I_D/I_M was larger in ethyl acetate solution relative to tetrahydrofuran for the (2N)methyl, (2N)propyl, and 2-(2N)ethyl polymers [154]. Finally, the 2-naphthylacrylate polymer and the 2-naphthyl methacrylate polymer are subject to a photo-Fries reaction [152].

Despite the similarity of the 2-naphthyl and 1-naphthyl polymers, several reports suggest that the detailed fluorescence behavior of the naphthylalkyl methacrylate polymers varies with the naphthyl substitution position and substituent chain length. The fluorescence lifetimes of monochromophoric model compounds for the 1-naphthyl [152], 1-naphthylmethyl [151], and 2-(1-naphthyl)ethyl [151] methacrylate polymers were found to be 10.2, 34, and 63 ns, respectively, in tetrahydrofuran at 298 K. The location of the oxygen atom relative to the naphthyl ring obviously affects the electronic properties (and chemical reactivity [152]) of the aromatic system. The fluorescence ratio of poly(2-naphthyl methacrylate) [55] was determined to be the same in both dichloromethane and ethyl acetate solution, in contrast to the corresponding 1-naphthyl polymer. It is unclear whether this is due to adjacent excimer formation in poly(2-naphthyl methacrylate), because there have been no studies on bis(2-naphthyl) model compounds for this or any other 2-naphthylakyl methacrylate polymer.

2.4.4 Summary

The review of the structure of excimers in aryl vinyl polymers can be concluded as follows:

1) In both intra- and intermolecular excimers, the aromatic rings are placed in a fully-eclipsed, parallel sandwich arrangement. This is true only for PS and P2VN; exceptions have been noted for PVK and P1VN.
2) Short-range intramolecular excimers (chromophores separated by 13 carbon atoms or less) are formed exclusively between nearest-neighbor rings (3 atom separation) in PS, P2VN, P1VN, and PVK.
3) Long-range intramolecular excimers (>13 atom separation) are not formed in PS, P2VN, P1VN, or PVK. This conclusion is based on the fluorescence behavior of Richards' polymers [143], head-to-head PS, and of alternating copolymers containing phenyl, 2-naphthyl, or *N*-carbazolyl chromophores. There is not enough data on the polymers of allylarenes or arylalkyl methacrylates to show conclusively whether long-range excimers are formed in these systems.
4) For PS and P2VN, intramolecular excimers are located largely at *meso* dyads in the tt conformation. The rotational states of the aromatic rings relative to the backbone must be consistent with the fully-eclipsed requirement for excimer formation. Only the low-energy excimer of PVK meets the above description. For P1VN, there may be additional complications relative to P2VN because of the energetics of the 1-naphthyl rotational states.
5) The number fraction q, of chromophores within intermolecular excimer-forming sites that are present in pure PS or P2VN cannot be accurately estimated. On one hand, the fluorescence of undiluted head-to-head PS, Richards' polymers [143], and alternating copolymers suggests that q is less than 10%. On the other hand, pure PBzMA exhibits considerable excimer fluorescence such that if energy migration is assumed to be nonexistent, q would be about 50%. This high EFS fraction would imply considerable order in amorphous PBzMA. The ideal experimental system for studying intermolecular excimer formation in P2VN would be a pure film of an alternating copolymer of styrene and 2-vinylnaphthalene, since this copolymer would have molecular packing very similar to that in pure P2VN. For the present, intermolecular EFS in rigid systems must be studied under conditions in which energy migration is active.

2.5 Properties of Excimers in Aryl Vinyl Polymers

2.5.1 Introduction

This section will be concerned with two basic questions on the properties of excimers in rigid systems containing aryl vinyl polymers:

— Does the fluorescence yield or lifetime of an intramolecular excimer differ from that of an intermolecular excimer?
— What are the effects of a rigid polymer matrix and of oxygen on the excimer fluorescence lifetime and quantum yield?

2.5.2 Comparison of Intermolecular and Intramolecular Excimer Formation in Model Compounds

The quantum yields and decay rates of the intermolecular excimer of naphthalene and its derivatives are given in Table 8. The solvent ethanol: water 95:5 v/v is one of the few solvents in which the fluorescence of these compounds has been completely characterized. Examination of the values of k_D and Q_M for other solvents shows that 95% EtOH does not belong in the same class as the hydrocarbon solvents, or even anhydrous ethanol. In the latter solvents, k_D/k_M falls between 0.8 for 1,6-dimethylnaphthalene and 1.4 for naphthalene. Although the quantity k_{FD}/k_{FM} has been measured only once for a naphthyl compound in a hydrocarbon solvent (sec Table 5), the values 0.3 and 0.4 seem appropriate for 1,6-dimethylnaphthalene and naphthalene, respectively, in hydrocarbon solvents. Since $Q_D/Q_M = (k_{FD}/k_{FM}) \div (k_D/k_M)$, we obtain $Q_D/Q_M = 0.4$ for 1,6-dimethylnaphthalene and 0.3 for naphthalene. The intrinsic quantum yield ratio as determined in 95% EtOH solvent is about seven

Table 8. Decay Rates and Intrinsic Quantum Yields of Naphthalene Compounds in Solution at Room Temperature

Compound	Solvent[b]	— (10^6 s^{-1}) —						
		k_D	k_M	k_D/k_M	Q_D	Q_M	Q_D/Q_M	Ref.
N[a]	95% EtOH	2.63	19.2	0.137	0.32	0.12	2.7	75)
	MeOH	—	11.8	—	—	0.23	—	156)
	T	12.5	9.1	1.37	—	—	—	71)
	CH	8.6	8.3	1.04	—	0.19	—	71)
	Hx	15	9.1	1.65	—	—	—	71)
	Paraffin	9.1	8.6	1.06	—		—	71)
1-methyl N	95% EtOII	3.3	24	0.138	0.34	0.12	2.8	75)
	EtOH	11.5	10.3	1.12	0.23	0.19	1.2	76)
	MeOH	—	14.9	—	—	0.21	—	156)
	T	15	12	1.25	—	0.18	—	71)
	CH	—	14.9	—	—	0.21	—	42)
	THF	—	18	—	—	—	—	45)
2-methyl N	95% EtOH	4.5	21	0.214	0.25	0.16	1.6	75)
	MeOH	—	16.9	—	—	0.28	—	156)
	CH	—	16.9	—	—	0.27	—	42)
	MCH	—	15.6	—	—	0.41	—	121)
2-ethyl N	DCM	—	105.	—	—	—	—	157)
	THF	—	18	—	—	0.27	—	12)
1,6-dimethyl N	95% EtOH	3.5	18	0.194	0.36	0.20	1.8	75)
	CII	18	19	0.95	—	0.28	—	71)
	PhCH	22	40	0.55	—	0.15	—	71)
	Hp	16	20	0.80	0.09	0.25	0.36	77)
2,6-dimethyl N	95% EtOH	—	23	—	0.3	0.30	1.0	75)
	CH	—	26	—	—	0.37	—	42)

[a] N = naphthalene.

[b] EtOH = ethanol; MeOH = methanol; CH = cyclohexane; T = toluene; THF = tetrahydrofuran; MCH = methylcyclohexane; DCM = dichloromethane; PhCH = phenyl cyclohexane; Hp = heptane; Hx = hexane.

times larger than that determined for hydrocarbon solvents. Because of this, the value $Q_D/Q_M = 0.4$ for 2-methylnaphthalene in a hydrocarbon solvent is a better estimate than the value of 1.6 given for 95% EtOH solvent in Table 8.

To determine whether Q_D and Q_D/Q_M for the intramolecular excimer differ from the corresponding intermolecular values, the fluorescence behavior of excimer-forming bis(2-naphthyl) compounds has been collected in Table 9. The experimentally-measured excimer and monomer quantum yields φ_D and φ_M are shown. Q_D was calculated in all cases by the relation $Q_D = \varphi_D/(1 - \varphi_M/Q_M)$, after assuming that Q_M for the bis(2-naphthyl) compound was 0.27, the same as for 2-methylnaphthalene.

For the intramolecular excimer of bis(2-naphthyl) compounds, the value $Q_D = 0.15 \pm 0.03$ is obtained in room-temperature solution; $Q_D/Q_M = 0.55 \pm 0.15$ is the quantum yield ratio corresponding to $Q_M = 0.27$. We note that $Q_D = 0.125$ was obtained for a P2VN sample having $M_w = 270{,}000$, for which $\varphi_D = 0.12$ and $\varphi_M = 0.010$ in 2-methyl THF solution [166]. Both Q_D and Q_D/Q_M for the intramolecular excimer fall within the range of values given for the intermolecular excimer in Table 8. Unfortunately, the solvent 95% EtOH has not been utilized with the bis(2-naphthyl) compounds, so a direct comparison with the 2-methylnaphthalene data is not possible.

Values of k_D which appear in Table 9 were obtained by the usual biexponential decay analysis [37] adapted from the intermolecular version due to Birks [71]. Despite observations made for *meso*-bis(1-(2-naphthyl)-ethyl) ether [13] and 1,3-bis(2-naphthyl) propane [159] of one rise time and *two* decay times in the transient fluorescence of the excimer, there have been no reports confirming this complication for the compounds

Table 9. Decay Rates[a] and Intrinsic Quantum Yields of Bis(2-naphthyl) Compounds in Solution at Room Temperature

Compound[b]	Solvent[c]				—(10^6 s^{-1})—		Ref.
		φ_D	φ_M	Q_D	k_D	k_{FD}	
1,3-bis (2N)propane	THF	0.13	0.033	0.16	12	1.9	12)
	CH	0.10	0.018	0.11	—	—	39)
	MCH	0.138	0.024	0.15	11.6	1.7	121)
	DCM	—	—	—	22	—	157)
1,3-bis (2N)butane	THF	0.14	0.026	0.16	12	1.9	12)
1,3-bis (2N)pentadecane	THF	0.14	0.025	0.16	12	1.9	12)
1,3-bis(2N)-5-phenylpentane	THF	0.138	0.022	0.15	15.3	2.3	12, 158)
meso-2,4-bis (2N)pentane	THF	0.14	0.020	0.16	12	1.9	12)
dl-2,4-bis (2N)pentane	THF	0.10	0.066	0.16	12	1.9	12)

[a] Obtained after analyzing the decay behavior by the biexponential scheme of Birks [71] and assuming that $k_M = 1.8 \times 10^7$ s^{-1} and $Q_M = 0.27$;
[b] 2N = 2-naphthyl;
[c] THF = tetrahydrofuran; CH = cyclohexane; MCH = methylcyclohexane; DCM = dichloromethane.

of Table 9. For all the bis(2-naphthyl) compounds, $k_D = 1.3 \pm 0.2 \times 10^7\ s^{-1}$, which leads to $k_D/k_M = 0.72$ if $k_M = 1.8 \times 10^7\ s^{-1}$ is assumed. Thus, the values of k_D for the intra- and intermolecular excimer are quite similar, if the intermolecular results taken in 95% EtOH are disregarded.

Finally, Table 9 shows that k_{FD} of the intramolecular excimer has the value $2.0 \pm 0.3 \times 10^6\ s^{-1}$ for all the bis(2-naphthyl) compounds. If k_{FM} is assumed to be $4.9 \times 10^6\ s^{-1}$, the same as for 2-ethylnaphthalene, then $k_{FD}/k_{FM} = 0.41$. Comparison of the intermolecular data in Table 5 with the intramolecular values for k_{FD} and k_{FD}/k_{FM} indicates the similarity of the two types of excimers. Also, the similarity in Q_D and k_D holds for the excimers of *meso* and *dl*-2,4-bis(2-naphthyl)pentane, compounds whose ground-state conformational behavior is quite different.

In summary, the intrinsic quantum yield and lifetime of the 2-naphthyl excimer is not significantly affected by the mode (intra vs. inter) of excimer formation, at least in a nonpolar solvent. The next question to be pursued is whether the similarity between intra- and intermolecular excimers is maintained in a rigid matrix.

2.5.3 Influence of Matrix Rigidity and Dissolved Oxygen

As noted earlier, the limiting lifetime of pyrene excimer fluorescence from concentrated solutions in PS and PMMA glasses was found to be the same as that of pyrene in cyclohexane solution. There have been no similar studies of naphthyl compounds in rigid glasses. Values of k_D and Q_D for the [2,6]-naphthalenophanes have not yet been determined for any solvent system. The bis(2-naphthyl) compounds have not been quantitatively characterized in rigid matrices, probably because excimer fluorescence is weak and difficult to detect under such conditions. Given such limited data, it can only be assumed that the values of Q_D and k_D of 2-naphthyl excimers remain the same in rigid solution as in fluid solution.

Excimer fluorescence from polychromophoric compounds in rigid systems, while easy to detect, is difficult to interpret. The transient response of the excimer can be empirically characterized by the limiting lifetime $\tau_{\infty D}$. In the absence of processes which convert D* to M*, this limiting lifetime is the reciprocal of k_D. We will examine $\tau_{\infty D}$ for PS, P2VN, and other aromatic polymers to see if there is any difference between fluid and rigid solution at room temperature.

For PS, the value of $\tau_{\infty D}$ in degassed solution has been found [4,37,160,161] to vary between 15.5 and 22 ns, the average being 18 ns. The rigid-solution value for bulk PS was first reported [37] to be 19 ns; a recent study [162] gives two equally-weighted excimer lifetimes of ~15 and 25 ns. The agreement of $\tau_{\infty D}$ for PS in solution with $\tau_{\infty D}$ for bulk PS supports the assumption that k_D is not affected by the rigidity of the solvent matrix.

The limiting lifetime of the P1VN excimer in cyclohexane [4] or dioxane [163] solution was given as 25–30 ns. Later studies in dichloromethane yielded $\tau_{\infty D} = 43$ ns [160] and 22 ns [130]. While bulk P1VN was not investigated, glassy films of PS or PMMA containing 0.5 wt % P1VN exhibited $\tau_{\infty D} \approx 50$ ns [151]. The fact that $\tau_{\infty D}$ is somewhat shorter in fluid solution than in rigid solution could be caused by the P1VN photodimerization [117] that occurs in solution but not in glasses.

For the 1-naphthylalkyl methacrylate polymers that do not photoreact, $\tau_{\infty D}$ has been found to be unaffected by the rigidity of the solvent matrix. Poly(1-naphthylmethyl methacrylate) has $\tau_{\infty D} = 48$ and 63 ns in THF and toluene, respectively, and $\tau_{\infty D} = 55$ ns in both PS and PMMA hosts [151]. Similarly, poly(2-(1-naphthyl)ethyl methacrylate) has $\tau_{\infty D} = 60$ and 76 ns in THF and toluene, respectively, and $\tau_{\infty D} = 73$ ns in a PS host [155].

The limiting lifetime of the P2VN excimer in dichloromethane [157] and THF [164] solution was given as 35 and 63 ns, respectively. The quenching properties of dichloromethane that shorten $\tau_{\infty D}$ have been previously noted in Table 9. For poly(2VN-*co*-S), $\tau_{\infty D} = 65$ ns in THF solution [158]. The THF solution values agree well with $\tau_{\infty D} = 66 \pm 4$ ns of glassy films of PS containing 0.3 wt % P2VN [165]. Thus, the value of $k_D = 1.5 \times 10^7\ s^{-1}$ given by the reciprocal of $\tau_{\infty D}$ does not depend on solvent fluidity. Moreover, k_D falls into the range of $k_D = 1.3 \pm 0.2 \times 10^7\ s^{-1}$ determined for bis(2-naphthyl) compounds in solution.

The final question to be considered is whether or not the oxygen in an air-saturated, glassy system quenches excimer fluorescence. Since the lifetimes of the phenyl and naphthyl excimers are the same order of magnitude as the lifetimes of the corresponding monomer species, and since the monomer fluorescence in glassy solvents is unaffected by the presence of air, we would expect that excimer fluorescence would not be affected by air. In fact, the excimer fluorescence of pure PS in a deoxygenated environment was found to decrease only 3% on the admission of air into the sample [36]. Experiments performed in this work on P2VN blend films held overnight in a nitrogen atmosphere and examined under nitrogen have detected no major changes in the P2VN fluorescence spectra relative to air-equilibrated samples. Similar results have been reported for PVK blend films [167].

In conclusion, the fluorescence yield and decay rate of both intra- and intermolecular excimers in P2VN in air-equilibrated glassy blends can be assigned single values, namely $Q_D = 0.15 \pm 0.03$ and $k_D = 1.3 \pm 0.2 \times 10^7\ s^{-1}$. These quantities are the same as for monomer bichromophoric 2-naphthyl compounds in deoxygenated hydrocarbon solvents, because neither the solvent rigidity nor the presence of oxygen in glassy systems affects Q_D and k_D.

3 Electronic Energy Migration and Bichromophoric Processes in Aryl Vinyl Polymers

3.1 Introduction

The study of electronic energy migration and transfer in organic glasses has a long history, starting in 1951 with the first plastic scintillators [168]. These early studies have been reviewed [169]. Later work focussed on the behavior of PS and PVK under the simpler conditions of excitation by UV light rather than high-energy particles. Klöpffer has written a series of reviews on these and other chromophor-bearing polymers in fluid solution [37], low-temperature rigid solution, [170] and in the pure state [37,171]. Electronic energy transfer in polymers has also been reviewed by Turro [172] and MacCallum [173].

This review section will cover recent experiments with PS and P2VN, which clearly show that there is some sort of singlet energy transfer from non-EFS chromophores to EFS chromophores. We will also briefly mention the studies of fluorescence and fluorescence quenching in solution, and of fluorescence depolarization in rigid media for various phenyl- and naphthyl-containing polymers and copolymers. Such studies are generally so complex and difficult to interpret that they do not help to decide whether singlet energy transfer occurs or not. The final section will deal with the nature of the energy transfer process from non-EFS to EFS chromophores. The Förster theory [177] will be utilized to estimate the rates of the possible migration and transfer steps, and to determine whether the process is single-step or multistep.

3.2 Evidence for Energy Migration in Poly(2-vinylnaphthalene) and Polystyrene

3.2.1 Fluorescence of Dilute Miscible Blends

Miscible polymer blends containing a small amount of an aryl vinyl polymer are ideal systems for studying singlet energy migration and transfer. First, the complication of excimer formation and dissociation in fluid solution caused by backbone rotations are avoided. In addition, by utilizing an air-saturated host matrix that is rigid at room temperature, all the triplet state kinetics can be simply included in the rate constant $k_{IM} = k_M - k_{FM}$. Finally, the number of EFS can be calculated in dilute miscible blends, because the EFS are limited to the adjacent intramolecular type. This limitation does not hold in an undiluted aryl vinyl polymer sample.

In order to determine whether energy migration makes a significant contribution to the photophysical behavior of P2VN and PS in dilute miscible blends, it is instructive to calculate the expected excimer-to-monomer fluorescence quantum yield ratio in the absence of energy migration. To do so, it is first necessary to assume that intermolecular and non-adjacent intramolecular EFS are absent. In addition, the adjacent intramolecular EFS are assumed to be frozen into the aryl vinyl polymer and must be excited by direct absorption of a photon. Since the absorption spectrum of an EFS is no different from that of non-EFS chromophores, then the calculated fraction of rings within EFS is sufficient to determine the fluorescence ratio.

For systems in which the triplet state decays quickly and radiationlessly to ground, the excimer-to-monomer fluorescence quantum yield ratio is given by

$$\varphi_D/\varphi_M = (Q_D/Q_M)\,(1/M - 1)\,, \tag{1}$$

where M is the probability that an absorbed photon ultimately decays from the monomer state. If there were no energy transfer in a rigid system, then

$$M = 1 - q\,, \tag{2}$$

where q is the number fraction of rings within EFS.

For a dilute miscible blend containing P2VN, we have shown that $Q_D/Q_M = 0.55 \pm 0.15$ and that $q \approx 2\,q_D = 0.026$ at room temperature. Substitution of these values into Eq. (1) along with the no-transfer expression for M, yields $\varphi_D/\varphi_M = 0.014$.

This is about 1/100th the experimental value of $I_D/I_M = 0.9$ recorded for 0.3% P2VN (70,000)/PS (2200) blends. [3c)] Thus, there must be some mechanism that increases the probability that an absorbed photon decays from the excimer state, since there are not enough intramolecular EFS to cause such a large value of I_D/I_M by direct excitation alone. Similar observations were made for dilute miscible PS/Poly(vinyl methyl ether) blends [140)].

Several proposed explanations of this effect that do not involve singlet energy migration can be discounted. The process of multiple monomer fluorescence/reabsorption, which would occur until the EFS finally absorbs the photon, would also seriously attenuate the high energy bands in the monomer fluorescence envelope. This distortion is not observed. Another proposal assumes that there is sufficient mobility, even in a glassy host matrix, to allow excimer formation through backbone rotations alone. However, we have shown that I_D/I_M is approximately 10^{-2} for 1,3-bis(2-naphthyl)propane in a PS (2200) host,[2h], which effectively negates this proposal. Finally, it has been suggested that excimers are formed by triplet-triplet annihilation at EFS, following a period of triplet energy migration. The insensitivity of the fluorescence ratio of room-temperature films to oxygen and to the excitation intensity rules out this possibility. In summary, there must be a single-step or multi-step mechanism whereby singlet-state energy is transferred radiationlessly from chromophores at the point of UV absorption to EFS chromophores.

3.2.2 Comparison of Polymer and Bichromophoric Model Compound Fluorescence in Dilute Solution

If singlet energy migration and transfer occurs for aryl vinyl polymers in dilute miscible blends, the same is also likely to result in dilute fluid solution. To demonstrate this it is necessary to show that the observed I_D/I_M values P2VN and PS in solution are larger than those estimated for these polymers in the absence of energy migration. This comparison may be made most effectively with reference to the behavior of model compounds in which no energy migration is possible.

For excimer-forming bichromophoric compounds in solution, the probability M of ultimate monomer decay in Eq. (1) is given by [37)]

$$M = [1 + (k'_{DM}/k_M)/(1 + k'_{MD}/k_D)]^{-1} . \qquad (3)$$

The excimer formation and dissociation rates in the above are primed, to show that they are due to backbone rotations and to distinguish them from the rigid-matrix processes described in Table 4. Each chromophore has only one neighbor, and the values of M for the two chromophores are identical. Energy migration cannot increase the fluorescence ratio of bichromophoric compounds, since there is only one dyad.

If energy migration were ignored, the fluorescence kinetics of aryl vinyl polymers in solution could be expressed by the same set of rate constants as utilized for the *meso-* and *dl*-2,4-diarylpentanes. However, each chromophore in the polymer (ignoring chain ends) has two neighbors which may participate in excimer formation. A chromophore centered in a heterotactic triad, for example, would generate excimers on the syndiotactic side and on the isotactic side at the rates $k'_{DM,dl}$ and $k'_{DM,meso}$, respectively. Excited monomer due to excimer dissociation within the heterotactic triad would appear at the central chromophore at a rate of $\frac{1}{2}(k'_{MD,dl} + k'_{MD,meso})$. An expression for M for an atactic aryl vinyl polymer generally cannot be written in closed form, because the nonzero values of k'_{MD} permit a make-and-break form of energy migration. However, if $k'_{MD}/k_D \ll 1$, then M can be calculated from Eq. (3) for the isotactic, heterotactic, and syndiotactic triads by taking $k'_{DM} = 2\,k'_{DM,meso}$; $k'_{DM,meso} + k'_{DM,dl}$; and $2k'_{DM,dl}$, respectively. The value of M for the polymer is then obtained by averaging over the triad population of the polymer.

The general case of nonzero k'_{MD} can be approximately treated by averaging M over the triad population. The probability M for the isotactic, syndiotactic, and heterotactic triads is taken equal to that for a polymer which is purely isotactic, purely syndiotactic, and alternately iso- and syndiotactic, respectively. It can be shown that M is given exactly for a purely isotactic polymer by Eq. (3), provided that $k'_{MD} = k'_{MD,meso}$ and that $k'_{DM} = 2$. A purely syndiotactic polymer can be treated in a similar way by utilizing rates from the *dl* isomer. A strictly alternating syndiotactic-isotactic polymer has the following exact expression for M:

$$M = [1 + (k'_{DM,meso}/k_M)/(1 + k'_{MD,meso}/k_D) + (k'_{DM,dl}/k_M)/(1 + k'_{MD,dl}/k_D)]^{-1} . \tag{4}$$

Thus, Eqs. (1)–(4) and the stated substitutions provide the means to approximate the fluorescence ratio of aryl vinyl polymers in solution, in the absence of energy migration.

The rate parameters for the model compounds of PS and P2VN are given in Table 10. The values of M for isotactic, syndiotactic, and heterotactic triads of P2VN can be calculated as 0.035, 0.147, and 0.0565, respectively. For the same triads of PS, the values of M are 0.0097, 0.172, and 0.0184, respectively. If we assume that a typical atactic polymer has 50% isotactic dyads, and if the dyads are independently distributed on the polymer, then there will be 25% isotactic and 25% syndiotactic triads. Thus, the value of M for a 50% isotactic P2VN sample in solution should be about 0.074; that for a 50% isotactic PS sample in solution should be 0.0545.

The value of φ_D/φ_M for P2VN relative to the 2-naphthyl compounds is calculated to have the proportion 4.8:4.3:1 for *meso*-2,4-bis(2-naphthyl) pentane, atactic

Table 10. Rate Parameters of 2,4-Diarylpentanes in Solution at Room Temperature

Aryl group/Isomer	Solvent[a]	—(10^6 s^{-1})—					Ref.
		k_M	k_D	k'_{DM}	k'_{MD}	M^{-1}	
phenyl/*meso*	IO	36[b]	40	2,000	4.0	52	135)
phenyl/*dl*	IO	36[b]	47	90	1.5	3.4	135)
2-naphthyl/*meso*	THF	18[c]	12	790	26	14.8	12)
2-naphthyl/*dl*	THF	18[c]	12	87	8.0	3.9	12)

[a] IO = isooctane; THF = tetrahydrofuran;
[b] Monochromophoric model compound = isopropylbenzene;
[c] Monochromophoric model compound = 2-ethylnaphthalene

P2VN, and the *dl*-pentane compound, respectively, assuming no energy migration. The experimental values for these respective compounds [12,166] follow the proportion 4.8:8.0:1, so that the fluorescence ratio of P2VN is nearly twice as large as expected.

The calculated value of φ_D/φ_M for PS lies in the proportion 28:7.2: 1 for *meso*-2,4-diphenylpentane, atactic PS, and *dl*-2,4-diphenylpentane, respectively. The experimental values for these respective compounds [17,37,135] fall into two conflicting proportions, because of a wide range in the fluorescence ratio of atactic PS. The proportion 28:6:1 reflects $\varphi_D/\varphi_M = 4$ for PS [37], while a recent determination of $\varphi_M = 17$ for PS [17] yields the proportion 28:24:1. The data for PS are inconclusive, but it will be shown later that energy migration in PS is probably slower and less effective than in P2VN.

We conclude that the difference between the experimental value and the no-transfer value of the fluorescence ratio of P2VN and PS is less in solution than in dilute miscible blends, because energy migration must compete with rotational processes in the generation of excimers in solution. This difference is also present when the effect of molecular weight on φ_D/φ_M of aryl vinyl polymers in solution and in dilute miscible blends is considered in the next section.

3.2.3 Influence of Molecular Weight

When the molecular weight of PS in dilute miscible PS/Poly(vinyl methyl ether) blends is lowered from 390,000 to 2200, the fluorescence ratio is observed [140] to drop to one-third of the high mol.-wt. value. Similar results were obtained [165] for dilute P2VN/PS blends in which the P2VN molecular weight was varied from 90,000 to 5500. The miscibility of these blends was not known, however. I_D/I_M for the miscible P2VN/PS (2200) blends examined recently [28] dropped from 1.4 to 0.58 for the P2VN (265,000) and P2VN (21,000) samples, respectively. This molecular-weight effect is inexplicable if no energy migration is assumed, because the fluorescence ratio depends only on the fraction of rings within EFS in the absence of energy migration. The number of adjacent intramolecular EFS does not depend on molecular weight, however [140].

In solution, the two end chromophores of a polymer chain produce a molecular-weight effect even without energy migration, because the end rings have only one neighbor and thus yield fewer excimers than the middle chromophores. The probability M for an aryl vinyl polymer with N repeat units is given by

$$M = (2/N)M_{end} + (1 - 2/N)M_{\infty} , \tag{5}$$

where M_{∞} is the value (previously calculated) for an infinite molecular-weight polymer of a given tacticity, and M_{end} is the value for the chain-end chromophores, again considering the tacticity of the polymer. This chain-end probability is given by the average of the polymer of the M_{meso} and M_{dl} values over the dyad population.

The fluorescence ratio of samples of P2VN and PS that are 50% isotactic and have $N = 10$ will be computed and compared with the infinite molecular-weight value, assuming no energy migration. First, M_{end} for P2VN and PS chain ends is calculated from the data in Table 10 to be 0.162 and 0.157, respectively. Second, recall that

M_∞ for atactic P2VN and PS samples was found to be 0.074 and 0.0545, respectively. Third, the value of M for P2VN and PS samples having N = 10 is given by Eq. (5) as 0.092 and 0.075. Finally, we compute from Eq. (1) that the value of φ_D/φ_M for the N = 10 polymer relative to the infinite molecular-weight value is 0.79:1 and 0.71:1 for P2VN and PS, respectively. Similar calculations for the N = 100 polymers show that the fluorescence ratios are within 96% of their infinite molecular-weight values.

Experimentally, the molecular-weight effect on φ_D/φ_M of P2VN [141,166], P1VN [174] and PS [25,175,176] in solution is larger and occurs over a broader range of molecular weights than calculated by the no-transfer approximation. From the previous discussion in Section 2.4 it is clear that these polymers do not form non-adjacent intramolecular excimer, so the observed molecular-weight effect cannot be explained in these terms. The process of energy migration, however, satisfactorily explains both the magnitude and the form of the dependence of the fluorescence ratio on the polymer molecular weight.

3.3 Fluorescence of Random Copolymers

A wide variety of copolymers of vinyl arenes with spectroscopically inert monomers have been synthesized since the first study of P(S-co-MMA) in 1963 [20]. The fluorescence of copolymers derived from styrene [178–180,184], 1-vinyl naphthalene [31,179,181–183], and 2-vinylnaphthalene [41,158,164,179] has been studied, primarily in fluid solution. Numerous functions relating φ_D, φ_M, and the fluorescence ratio to the copolymer composition have been proposed, none of which can be applied over the entire composition range of the copolymer.

Given the three types of triads ARA, ARR and RRR which are present in copolymers (R = vinyl arene residue, A = inert monomer residue), it is not surprising that the fluorescence behavior of copolymers in solution is complex. The probability of ultimate monomer decay M of the center chromophore in each type of triad is different, and the ring pair in AAR may be iso- or syndiotactic. The ring triplet in RRR, moreover, may be iso-, syndio-, or heterotactic. Thus, even in the absence of energy migration, the average value of M must reflect the triad composition of the copolymer and the tacticity of the ARR and RRR sequences. The fluorescence of copolymers containing methyl methacrylate comonomer may be further complicated by intramolecular quenching of the chromophores by the carbonyl moiety (see Sect. 2.2).

Recently, an analysis in terms of the three copolymer triads of φ_D and φ_M of P(2VN-*co*-S) copolymers in solution was made [158]. The experimental data were successfully fitted only after additional energy migration terms, which allow the transformation AR*A to AR*R or RR*R, were added. The fluorescence decay curves of P(1VN-*co*-MMA) [44] and P(1VN-*co*-methyl acrylate) [45] in solution were interpreted by a somewhat simpler scheme, in that the triads ARR and RRR were lumped together. Nevertheless, an energy migration term linking AR*A to AR*R or RR*R was required in order to yield the observed decay behavior. These results suggest that energy migration has sufficient range to "jump" over short sequences of the inert monomer A in the copolymer. Additional work on φ_D and φ_M

of well-characterized copolymers and of trimeric model compounds [158] for the copolymer must be done if the photophysics of copolymers in solution are to be developed further, and if definite conclusions on the nature of energy migration in copolymers are to be drawn.

3.4 Fluorescence Depolarization

The technique of fluorescence depolarization is useful for studying energy migration and transfer in rigid systems, because a photon from a polarized light source that initially excites a chromophore will lose that polarization during the energy migration process. When a sample is excited by light polarized in the vertical plane and traveling horizontally, the polarization p of the resultant fluorescence is defined by

$$p = (I_{||} - I_{\perp})/(I_{||} + I_{\perp}) . \tag{6}$$

The intensities $I_{||}$ and $I_{\perp}$ are both observed in a direction normal to the plane of polarization of the exciting light. $I_{||}$ and $I_{\perp}$ denote intensities passed by vertically- and horizontally-aligned polarizers, respectively.

There are numerous problems which must be resolved before the results of a fluorescence depolarization experiment can be interpreted. The orientation of the absorbing chromophores, if nonrandom, must be known before calculating the polarization of the sample immediately after excitation. The angle α between the absorption and emission dipoles within a chromophore must be known, since this angle induces depolarization of the photon at each energy migration step and at the fluorescence emission step. To date, α has not been measured for excimers in rigid systems. Finally, the technical problems of depolarization by light scattering from imperfections in low-temperature solvent glasses and in polymer films must be overcome.

Despite these difficulties, the depolarization of monomer fluorescence of copolymers of styrene [32, 179], 1-vinylnaphthalene [32, 179, 183, 185], and 2-vinylnaphthalene [179] has been studied in low-temperature solvent glasses. The same experiments were performed on PS [32, 186] and P1VN [32, 186]. Monomer fluorescence depolarization from room-temperature films of styrene [187] and 1-vinylnaphthalene [187] copolymers, and from blends of P(1VN-co-MMA) with PMMA [179], has also been recorded. Finally, excimer fluorescence depolarization from room-temperature films of PS [188, 189] and P(alpha-methyl S) [189] has been studied.

Several generalizations can be drawn from these experiments:

1) The polarization of copolymers containing 1% or less of the aromatic monomer is roughly the same as that of the monochromophoric model compound. In both solvent glasses and room-temperature films, $p = 0.1$ to 0.2.
2) For copolymers containing 80% or more of the aromatic monomer, $p \leq 0.03$ in low-temperature glasses or in dilute room-temperature blends. Note that excimer fluorescence is virtually absent in the former case.
3) Conflicting results have been obtained for the monomer emission of PS and P1VN in low-temperature glasses. David et al. [32] have measured $p = 0.03$ and 0.005 for PS and P1VN, respectively, while MacCallum [186] has found $p = 0.09$ and 0.33,

respectively. The latter two values seem suspect, given the copolymer data in point (2). Unfortunately, neither group seems to have utilized any sort of polarization standard.

4) Few conclusions can be drawn for room-temperature films of copolymers or polymers. David et al. [187] measured $p \leqq 0.02$ for copolymers containing as little as 35% of the aromatic monomer. In contrast, MacCallum et al. [188, 189] found that $p \approx 0.2$ for the excimer emission of PS and P(α-methyl S). The latter may be consistent with limited energy migration in pure PS, because the fraction of EFS chromophores in pure PS has recently been estimated to be about 40% [148].

In summary, the energy migration implied by monomer fluorescence depolarization of copolymers having a high content of aromatic units is most predominant under conditions where the number of EFS is small, i.e. in low-temperature solvent glasses. Whether the energy migration occurs solely by a singlet mechanism, or is enhanced by triplet migration and annihilation in these low-temperature glasses, is largely unknown. Much remains to be done on the basic theory of fluorescence depolarization of polychromophoric compounds, which is linked with the larger problem of the photophysics of aryl vinyl polymers.

3.5 Fluorescence Quenching

3.5.1 Excimer Forming Polymers

Quencher exciplexes are similar to excimers in that both can occur in fluid and rigid solvents [73]. Quenching occurs in rigid solvents because any of the chromophores surrounding a quencher molecule Q will, upon excitation, immediately form an exciplex having a negligible fluorescence quantum yield. While such static quenching also occurs in fluid solution, the predominant mechanism is the diffusional encounter of an excited chromophore and a quencher molecule. This dynamic quenching proceeds at the rate $k_Q[Q]$, where k_Q is viscosity-dependent. Singlet energy migration enhances quenching in rigid solvents, because energy is ultimately channeled into the quenched-chromophore sites. The enhancement is less in fluid solvents, however, since energy migration must compete with the rapid diffusion of the quencher.

Despite the advantages of studying fluorescence quenching in rigid, room-temperature matrices, there are technical problems with adding sufficient quencher without eliminating the rigidity of such systems. If oxygen is to be utilized as a quencher, then thin films of polymer samples can be kept under high-pressure oxygen during spectroscopic examination. This was done to quench the excimer fluorescence of pure PS films [36], in the only rigid-matrix quenching study of PS or P2VN to date. The excimer intensity was found to follow the relationship $I_{D,O}/I_D = 1 + 0.147[O_2] + 1.8 \times 10^{-3}[O_2]^2$, where $[O_2]$ is given in atm. If all the chromophores in pure PS were assumed to be within EFS, then the Perrin active-sphere quenching model [73] and the quencher-complex model [73] would predict that $I_{D,O}/I_D = e^{k[O_2]}$ and $1 + K[O_2]$, respectively. Neither model accurately represents the empirical curve, because the undisputed presence of non-EFS chromophores in PS complicates the analysis of excimer fluorescence quenching.

Almost all studies of fluorescence quenching of aryl vinyl polymers have been made in fluid solution, since the technique is simpler than rigid-matrix quenching. However, the kinetics of fluid-solution quenching, like the fluorescence kinetics of unquenched polymers, have not been satisfactorily described. The simple fluorescence scheme conceived for bichromophoric compounds [37] that was utilized in Section 3.2 to estimate φ_D/φ_M for aryl vinyl polymers in the absence of energy migration gives quenching relationships that are generally more complex than the Stern-Volmer type [73].

If dynamic quenching in which the quencher exciplex does not regenerate excited monomer is assumed, then the quencher Q reduces the fluorescence intensity according to the following equations:

$$I_{D,O}/I_D = [(1 + K_{QD}[Q])(1 + K_{QM}[Q]) - Z]/(1 - Z) \quad (7)$$

$$I_{M,O}/I_M = [1 + K_{QM}[Q] - Z/(1 + K_{QD}[Q])]/(1 - Z) \quad (8)$$

$$(I_D/I_M)_O/(I_D/I_M) = 1 + K_{QD}[Q], \quad (9)$$

where $K_{QD} = k_{QD}/(k_D + k'_{MD})$, $K_{QM} = k_{QM}/(k_M + k'_{DM})$, $Z^{-1} = (k_D + k'_{MD})(k_M + k'_{DM})/(k'_{MD}k'_{DM})$, k_{QD} and k_{QM} are the rates of dynamic quenching of excimer and monomer, respectively, and the other rates have been defined previously. Note that when $k'_{MD} \ll k_D$, then $Z \approx 0$ and Eqs. (7) and (8) simplify considerably.

Fluorescence quenching of PS [17,161,190,191], P2VN [157,192–194], poly(1-naphthyl methacrylate) [149,199], poly(2-naphthylmethyl methacrylate [155], P(S-*co*-MMA) [161], and P(2VN-*co*-S) [194] in solution has been examined. In all cases except one for PS [161], plots of $I_{D,O}/I_D$ versus [Q] showed upward curvature similar to Eq. (7). Excimer fluorescence was quenched more rapidly than monomer fluorescence with increasing [Q]. In addition, the observed quenching of the fluorescence ratio and of the monomer fluorescence was roughly linear in [Q].

These results are usually interpreted by the following procedure. First, Z is assumed to be close to zero, so that the observed biexponential decay rates λ_1 and λ_2 are equal to $k_M + k'_{DM}$ and k_D, respectively. [71] Second, K_{QM} for the polymer is obtained from Eq. (8) and k_{QM} is given by $K_{QM}\lambda_1$. Third, $K_{QM}k_M$ for a copolymer containing a small amount of the aromatic monomer of interest is measured, or else $1/2K_{QM}k_M$ for the monochromophoric model compound is determined. Either quantity is an estimate of k_{QM} in the aryl vinyl polymer, assuming no energy migration. Finally, k_{QM} of the polymer is compared with the estimate provided by the copolymer or model compound. If the polymer has a larger value of k_{QM} than the estimated value, then this implies that energy migration occurs in the polymer.

Because the monomer fluorescence intensity is difficult to extract from the spectra of PS, P2VN and other polymers having large fluorescence ratios, the interpretation of fluorescence quenching data varies widely. For PS, k_{QM} has been found to be 1.5 [190] or 6 [17] times as large as half the value of k_{QM} for ethylbenzene, signifying energy migration. The latter value [17] is probably too high, because it is based on $k_M + k'_{DM} = 4 \times 10^9\ s^{-1}$; others [160,190] have measured $k_M + k'_{DM}$ based on the biexponential decay scheme and have obtained $1.4–0.55 \times 10^9\ s^{-1}$. In contrast, energy migration in PS has been discounted by MacCallum et al. [161], who compared k_{QD} for PS with k_{QM} for a styrene-methyl methacrylate copolymer having less than 10% styrene links. It was claimed that since k_{QD} was not larger than k_{QM}, then no energy migration occurred in PS. This comparison is in error, however, since it is k_{QM} and not k_{QD} for PS which must be compared with k_{QM} for the styrene copolymer.

The interpretation of the fluorescence quenching of P2VN in solution has led to conflicting conclusions on energy migration. The value of k_{QM} for P2VN has been determined [194] to be four times as large as k_{QM} for a 2-vinylnaphthalene-styrene copolymer containing 1% 2-vinylnaphthalene links. In contrast, k_{QM} for P2VN was only about 45% of the quantity $1/2k_{QM}$ for 2-ethylnaphthalene [157]. There are two discrepancies in the latter report which have artificially increased k_{QM} for 2-ethylnaphthalene and decreased k_{QM} for P2VN, however. First, k_M for 2-ethylnaphthalene was measured [157] as 1.05×10^8 s^{-1} in aerated dichloromethane. While this solvent is known to increase k_M for aromatic compounds ($k_{M,CH_2Cl_2}/k_{M,THF} = 2.6$ for ethylbenzene [43]), the values of k_M in Table 8 are less than one-fifth the supposed dichloromethane-solvent value. Given that $k_M = 1.8 \times 10^7$ s^{-1} for 2-ethylnaphthalene in THF, [12] we expect that $k_M \leqq 5 \times 10^7$ s^{-1} for 2-ethylnaphthalene in dichloromethane. Second, $k_M + k'_{DM}$ for P2VN was found [157] to be 1.35×10^8 s^{-1}, while the value 3.0×10^8 s^{-1} was recorded [157] for 1,3-bis(2-naphthyl) propane in the same solvent. However, k'_{DM} for aryl vinyl polymers is usually greater than or equal to k'_{DM} for biochromophoric model compounds. Moreover, a recent study of P2VN in solution [195] has found that the major decay component within the monomer fluorescence response curve measures 4.8×10^8 s^{-1}. Thus, we expect that $k_M + k'_{DM} \geqq 3 \times 10^8$ s^{-1} for P2VN in dichloromethane. If the suggested changes in the decay rates are made, then k_{QM} for P2VN becomes about twice as large as the quantity $1/2k_{QM}$ for 2-ethylnaphthalene, reversing the earlier conclusion that there was no energy migration in P2VN [157].

Fluorescence quenching has been exhaustively investigated for poly(1-naphthyl methacrylate) [149,199]. From the polymer intrinsic viscosity, the solvents $CHCl_3$, THF, and benzene were selected to vary the coil size from large to medium, respectively [150,199]. Quenching rates k_{QM} were determined for the polymer and the model compound, 1-naphthyl acetate, in all three solvents. The ratio $k_{QM,polymer}/1/2k_{QM,model}$ was given as 1.7, 1.9, and 4.9, with respect to the solvents listed above [199]. It was concluded [149,199] that energy migration occurs in poly(1-naphthyl methacrylate) in solution, particularly so in poor solvents [199]. A final point in favor of energy migration can be taken from the quenching behavior [155] of poly(2-naphthylmethyl methacrylate), an excimer-forming polymer. The monomer quenching rate k_{QM} of this polymer in aerated 2-methyl THF was 1.8 times as large as k_{QM} of a 2-naphthylmethyl methacrylate-methyl methacrylate copolymer having 4% of the 2-naphthyl chromophore [155].

The analysis of fluorescence quenching in excimer-forming polymers embodied by Eqs. (7)–(9) contains a number of unresolved flaws. First, the decay schemes of P2VN [11,195] and other polymers [44,45,151,158,195] in solution require at least three decay terms to adequately interpret the fluorescence behavior over a range of temperatures. While additional work on the decay response of polymers as a function of [Q] is needed, the current quenching scheme is probably too simplistic. Second, an

EFS-like state that can form excimer in less than the backbone rotation time k'^{-1}_{DM} and is present in amounts much larger than expected from conformational statistics [126,140,141] has been postulated for PS in solution [17,190]. Additional decay studies in the subnanosecond range should help clarify this process for PS and P2VN. Finally, $k'_{MD}/k_D \approx 0.7$ has been measured for bis(2-naphthyl) model compounds in solution, [12] which suggests that $Z \approx 0.4$ for P2VN. Thus, the value of k_{QM} for P2VN obtained from Eq. (8) by assuming $Z = 0$ could be too large. Unfortunately, a fluorescence decay analysis including k'_{MD} does not appear in most fluorescence quenching studies.

3.5.2 Non Excimer-Forming Polymers

The ideal polymer for studying enhanced monomer quenching k_{QM} due to singlet energy migration would be one in which excimer formation is absent. Unfortunately, excimers cannot be excluded from the aryl vinyl polymers unless low temperatures are employed, and these conditions favor long triplet lifetimes, triplet migration and triplet-triplet annihilation. However, there are certain types of aromatic polymers in which excimer formation is absent at all temperatures. Quenching of monomer fluorescence in these polymers will be reviewed next.

There are a variety of arylalkyl methacrylate homopolymers which do not exhibit excimer fluorescence in dilute solution. Such polymers containing the phenyl [43,147], carbazolyl [196,197], or phenanthryl [197] chromophore have been examined. Fluorescence quenching has been studied only in a series of phenyl-containing methacrylate polymers [43], however.

Quenching rates k_{QM} were determined [43] utilizing CCl_4 for the homopolymers of benzyl, 2-phenylethyl, and 3-phenylpropyl methacrylate. Similarly, k_{QM} were measured for copolymers of methyl methacrylate containing $<10\%$ of each of the aromatic monomers. When k_{QM} of the homopolymers given above are divided by k_{QM} of the matching copolymer, the values 0.95, 1.13, and 1.81, respectively, are obtained for dichloromethane solvent. Values of the same quantity determined in ethyl acetate solvent are 1.38, 1.09, and 1.83, with respect to the phenylalkyl units listed above.

The fact that $k_{QM,\,polymer} > k_{QM,\,copolymer}$, seen especially for poly(3-phenylpropyl methacrylate) and for poly(benzyl methacrylate) in ethyl acetate, suggests that energy migration occurs in these polymers. However, it is assumed that the methyl methacrylate segments in the copolymer coil provide the same local environment as in the homopolymer coil. For example, k_{QM} for the copolymers were found [43] to be 0.7 to 1.5 times as large as $1/2\,k_{QM}$ for monochromophoric model compounds such as benzyl acetate, an effect attributable to the screening and/or retentive properties of the copolymer coil. Unfortunately, the intrinsic viscosity of the phenylalkyl methacrylate polymers was not determined in the solvents utilized. If poly(benzyl methacrylate) behaves like poly(1-naphthyl methacrylate) [199] in solution, then the enhanced value of k_{QM} for the former in ethyl acetate relative to CH_2Cl_2 can be attributed to increased energy migration in the coil-contracting solvent, ethyl acetate.

The quenching properties of poly(1-(2-naphthyl)ethyl methacrylate) provide an interesting comparison with those of poly(2-naphthylmethyl methacrylate), because the excimer fluorescence of the former is greatly reduced relative to the latter [155]. It was found that k_{QM} of the 1-(2-naphthyl)ethyl methacrylate polymer was only 1.3 times as large as k_{QM} of a methyl methacrylate copolymer containing 3% of the aromatic monomer [155]. Under identical experimental conditions, the 2-naphthylmethyl methacrylate polymer had $k_{QM} = 1.8\ k_{QM,\,copolymer}$ [155]. The apparently higher level of energy migration in the latter polymer may be due to the structural differences between the polymers, but more likely is an artifact of the approximation that $Z = 0$ in the 2-naphthylmethyl methacrylate polymer. Nevertheless, energy migration occurs at a moderate rate in both polymers.

3.5.3 Summary

In conclusion, fluorescence quenching reveals the following features of the fluorescence of aromatic polymers:

1) The quenching rate of non-excimeric arylalkyl methacrylate polymers in solution is greater than would be expected in the absence of energy migration. The increase is not totally due to retention of the quencher within the polymer coil, because the quenching rate of the same chromophore dispersed at low concentration in a copolymer is not as large as for the homopolymer.
2) The apparent quenching rate of monomer fluorescence of PS, P2VN, and other excimer-forming polymers in solution is greater than would be expected in the absence of energy migration. Moreover, the observed/expected quenching ratio is larger for aryl vinyl polymers ($n = 3$) than that for the non-excimeric arylalkyl methacrylate polymers ($n \geqq 7$). This is in accord with a lower energy migration rate in the latter polymers, due to the larger chromophore separation.
3) Quenching rates derived for excimer-forming polymers are questionable, given that the decay kinetics of unquenched samples are uncertain. More reliable values of k_{QM} can be obtained for non-excimeric polymers.

3.6 Energy Migration and the Förster Mechanism

The existence of radiationless singlet energy transfer between two dissimilar chromophores was established about 30 years ago in the classic experiments of Bowen [200–202]. A review of the field up to 1973 containing tables of the energy transfer properties of 210 compounds has been given by Berlman [203]. According to the theory of Förster [177,204], the dominant electrostatic interaction that causes radiationless transfer between two chromophores is dipole-dipole resonance. This yields energy transfer between a donor "d" and acceptor "a" at the rate

$$k_{d \to a} = k_{M,d}(R_0/R)^6 \,, \tag{10}$$

where $k_{M,d}$ is the decay rate of the donor in the absence of energy transfer, R is the distance between the donor and acceptor, and R_0 is the Förster radius [177]. R_0 is proportional to the overlap of the fluorescence spectrum of the donor and the absorption spectrum of the acceptor, and also depends on the relative orientation of the donor emission and acceptor absorption dipoles. Values of R_0 are commonly tabulated [203] under the assumption [177] that the donor and acceptor molecules are translationally fixed and randomly oriented, but rotationally free within the lifetime of the transfer process. While the detailed averaging of the dipole orientations in static systems can be quite involved [205], the primary factor in determining $k_{d \to a}$ is R, the donor-acceptor separation.

In excimer-forming polymers, the acceptors are the EFS and the donors are the monomer species. Because the number of EFS are difficult to estimate and do not remain fixed under various experimental conditions, acceptors have been deliberately introduced as in-chain impurities into various aromatic polymers. The fluorescence properties of 1-naphthylalkyl methacrylate polymers with anthryl tags [206–210], P2VN with pyrenyl tags [211], and 9-phenanthrylmethyl methacrylate polymers with anthryl tags [198] have been examined. Values of R_0 for 1-methylnaphthalene/9-methylanthracene, 2-methylnaphthalene/pyrene, and phenanthrene/9-methylanthracene are 23, 31, and 23 Å, respectively [203]. These Förster radii, while relatively large, do not encompass all the donor chromophores since the polymers are typically tagged with 0.1–1 mol % acceptor. Thus, it is likely that some energy migration occurs among the donors prior to the transfer step to the acceptors [208]. To further complicate matters, all except the phenanthryl polymers [198] emit excimer fluorescence even when tagged, and the overlap of the acceptor and excimer fluorescence bands is difficult to resolve.

Despite these problems, the ratio of the acceptor to the donor fluorescence quantum yields has been measured for four different polymer/tag systems in fluid solution. The probability that the donor singlet energy is transferred to the acceptor in systems containing 0.14–0.2 mol % acceptor was found to be 21, 10, 9, and 5% for poly(9-phenanthrylmethyl methacrylate)/A [198], poly(1-naphthylmethyl methacrylate)/A [209], poly(2-(1-naphthyl)ethyl methacrylate)/A [209], and P2VN/pyrene [211], respectively (A = anthracene tag). Thus, the transfer efficiency decreases as the amount of excimer fluorescence increases. Moreover, I_D/I_M of the tagged polymers remains virtually the same as that of the untagged polymers. We conclude that excimers and acceptors both compete for the energy of the monomer donor.

The donor-acceptor transfer probability has been reported to increase with decreasing solvent quality for anthryl-tagged poly(1-naphthyl methacrylate) [208]. In contrast, no solvent dependence of the transfer probability was seen for pyrene-tagged P2VN [211], probably because P2VN expands less than the methacrylate polymers in good solvents. The amount of acceptor fluorescence increases considerably in the spectra of neat films of the tagged polymers, and to a lesser extent for dilute blends of the tagged polymers [209–211]. It is not surprising that the fluorescence of adventitious in-chain acceptors in "pure" homopolymers exhibit many of the same characteristics as that of intentionally-tagged polymers, i.e. the effect of solvent quality [25,33] and of the undiluted polymer [25,128,145] on the amount of acceptor fluorescence.

The acceptors in untagged PS and P2VN are the EFS, which have essentially the same absorption spectrum as the monomer species (see Sect. 2.1 and 2.4). Thus, the value of R_0 for transfer from monomer to an EFS is the same or slightly larger than R_0 for migration from monomer to monomer. However, the overlap of the monomer fluorescence and absorption spectra is small, so that $R_0 = 6.47$ Å for isopropylbenzene/isopropylbenzene and $R_0 = 11.75$ Å for 2-methylnaphthalene/2-

methylnaphthalene [203]. The question that must be answered is whether the Förster mechanism predicts an energy migration rate that is large enough to yield the step or steps necessary for transfer to the EFS.

An approximate distance between nearest-neighbor rings in aryl vinyl polymers can be obtained from the cube root of the repeat-unit molar volume V of the undiluted polymer at room temperature. Using this approach, the nearest neighbor distances in PS and P2VN are found to be 5.5 and 6.1 Å, respectively. The rate of nearest-neighbor energy migration, k_{MM}, is given by Eq. (10) as a multiple of k_M, i.e. $k_{MM}/k_M = 2.7$ and 51 for PS and P2VN, respectively. The migration rate for second-nearest neighbors is roughly $(1/2)^6 = 1.6\%$ of the rate between nearest neighbors, so that the R^{-6} distance dependence of the Förster expression effectively limits a single migration step to nearest neighbors if the chain is in dilute solution. Similar conclusions were reached in a more sophisticated calculation of migration rates for P2VN based on the Förster mechanism [141].

Although migration over many chromophores in a single step is impossible, the same may be accomplished in a large number of nearest-neighbor steps. The average number of nearest-neighbor steps made in a polymer without EFS is simply $2\,k_{MM}/k_M \approx 5$ and 100 for PS and P2VN, respectively. Since the process is a random walk in one dimension, the number of distinct chromophores that are visited in n steps is proportional to $\sqrt{n}$ for n large. The range of the migration process can be estimated for P2VN as $(2\,k_{MM}/k_M)^{1/2} = 10$ chromophores. Significant energy migration to the EFS would be expected if the dyad fraction of EFS, q_D, equaled 1/10. However, it was shown earlier that $q_D = 0.013$ for P2VN at room temperature, and that there are substantial amounts of excimer fluorescence from dilute miscible blends containing P2VN. Thus, the Förster dipole-dipole estimate of the nearest-neighbor energy migration rate in P2VN and PS is too low.

There is an obvious error which occurs when the electrostatic interaction between two chromophores separated by only 6 Å is approximated by the dipole-dipole term. At such close distances, the chromophores cannot be approximated by point dipoles, and a variety of other short-range interactions that depend more strongly on R than R^{-6} must be entered into the theory [177]. To obtain an upper bound on k_{MM} under these conditions, the value of $k_{MM} \sim 10^{12}\ s^{-1}$ for crystalline naphthalene [212], in which the nearest-neighbor distance is 5.1 Å, can be taken. Next, the value of $k_M = 1.8 \times 10^7\ s^{-1}$ for 2-ethylnaphthalene gives $2\,k_{MM}/k_M \sim 10^5$, the average number of nearest-neighbor steps in the energy migration process. The range of the migrating singlet state thus has an upper limit of 300 chromophores in a 1D random walk in P2VN, which is more than adequate to account for the observed excimer fluorescence.

In summary, the Förster dipole-dipole energy migration rate provides a lower bound on k_{MM} in P2VN and other aryl vinyl polymers. The upper bound for k_{MM} was taken from migration rates in organic crystals. Because of the short range of the electrostatic interaction between like chromophores, all migration steps are assumed to occur between nearest neighbors such that the random walk of the singlet-state excitation is strictly one-dimensional. It is important to note that this will be true only for dilute systems; it will not be valid for the bulk polymer.

4 Conclusions

The objective of this review has been to characterize the excimer formation and energy migration processes in aryl vinyl polymers sufficiently well that the excimer probe may be used quantitatively to study polymer structure. One such area of application in which some measure of success has already been achieved is in the analysis of the thermodynamics of multicomponent systems and the kinetics of phase separation [2]. In the future, it is likely that the technique will also prove fruitful in the study of structural order in liquid crystalline polymers.

The review has touched upon the majority of all aromatic chromophore containing polymers but the emphasis has been on aryl vinyl polymers containing phenyl or naphthyl rings. Although this is clearly a restriction, it is justified by the specificity of information obtainable from systems containing these polymers. As the size of the aromatic ring increases, the variety of suitable excimer forming sites also increases. Consequently, there is a loss of detail in the knowledge of the nature of the environment surrounding the EFS traps.

Specific conclusions relevant to the photophysics of PS and P2VN are summarized below:

1) Ultraviolet quanta absorbed by the polymer are localized at individual chromophores, which have the same absorption spectra as monochromophoric model compounds.
2) The intrinsic quantum yield of monomer fluorescence Q_M for the polymer is the same as for the monochromophoric model compound. Q_M in air-equilibrated rigid hosts is the same as in degassed hydrocarbon solvents, and is independent of the host matrix in rigid systems.
3) Triplet lifetimes for the polymer in rigid, room-temperature hosts in air are very short due to O_2 quenching. Thus, processes like triplet-triplet annihilation that might affect the fluorescence quantum yield can be neglected.
4) EFS have the same UV absorption spectrum as a pair of non-EFS chromophores.
5) The fluorescence of phenyl or 2-naphthyl excimers is characterized by a single broad peak for both inter- and intramolecular formation. The parallel sandwich arrangement of chromophores is the only possible configuration for the excimer.
6) Intermolecular EFS and intramolecular EFS formed by rings separated by more than three backbone atoms can be neglected in dilute miscible blends. EFS are primarily located at *meso* dyads in the tt conformation.
7) The intrinsic quantum yield of excimer fluorescence Q_D for intermolecular excimers is the same as for intramolecular excimers. Q_D in air-equilibrated rigid hosts is the same as in degassed hydrocarbon solvents, and is independent of the host matrix in rigid systems.
8) In the guest polymer, there is a net radiationless transfer of singlet energy from chromophores to EFS. This must take place by a non-collisional process in rigid hosts. Förster dipole-dipole resonance theory indicates that single-step transfers over distances large relative to the repeat-unit distance are unlikely. However, the theory indicates that multistep migration among the chromophores of P2VN or PS can occur.

9) Singlet energy migration occurs in P2VN and PS via a random-walk process, based on the observation that the UV absorbance spectrum of the polymer is the same as that of a monochromophoric compound.
10) The rate of the transfer step from a non-EFS chromophore to the closest ring in a nearby EFS is the same as that of a migration step between two non-EFS rings separated by the same distance.

Acknowledgement: This work was supported by the Polymers Program of the National Science Foundation.

5 References

1. a. Anufrieva, E. V., Gotlib, Y. Y.: Advances in Polymer Science *40*, 1 (1981)
   b. Ghiggino, K. P., Roberts, A. J., Phillips, D.: ibid. *40*, 69 (1981)
2. a. Frank, C. W., Gashgari, M. A.: Macromolecules *12*, 163 (1979)
   b. Frank, C. W., Gashgari, M. A., Chutikamontham, P., Haverly, V., in: "Structure and Properties of Amorphous Polymers", Walton, A. G. (Ed.) Elsevier: Amsterdam 1980, pp. 187–210
   c. Semerak, S. N., Frank, C. W.: Macromolecules *14*, 443 (1981)
   d. Gashgari, M. A., Frank, C. W.: ibid. *14*, 1558 (1981)
   e. Frank, C. W., Gashgari, M. A.: Ann. N. Y. Acad. Sci. *366*, 387 (1981)
   f. Semerak, S. N., Frank, C. W.: Advances in Chemistry Series, *203*, 757 (1983)
   g. Semerak, S. N., Frank, C. W.: ibid., submitted
   h. Semerak, S. N., Frank, C. W.: Macromolecules, submitted
3. Birks, J. B.: "Photophysics of Aromatic Molecules", Wiley-Interscience: New York 1970; Chapter 2
4. Hirayama, F.: J. Chem. Phys. *42*, 3163 (1965)
5. Vala, M. T., Jr., Haebig, J., Rice, S. A.: ibid. *43*, 866 (1965)
6. a. Nguyen-Luong, B. V., Noël, C., Monnerie, L.: J. Polym. Sci., Polym. Symp. No. *52*, 283 (1975)
   b. Bokobza, L., Jasse, B., Monnerie, L.: Eur. Polym. J. *13*, 921 (1977)
7. Tatemitsu, H., Ogura, F., Nakagawa, Y., Nakagawa, M., Naemura, K., Nakazaki, M.: Bull. Chem. Soc. Jpn. *48*, 2473 (1975)
8. Stützel, B., Miyamoto, T., Cantow, H.-J.: Polym. J. *8*, 247 (1976)
9. Chandross, E. A., Dempster, C. J.: J. Am. Chem. Soc. *92*, 3586 (1970)
10. Chandross, E. A., Dempster, C. J.: ibid. *92*, 703 (1970)
11. Demeyer, K., van der Auweraer, M., Aerts, L., De Schryver, F. C.: J. Chim. Phys. *77*, 493 (1980)
12. Ito, S., Yamamoto, M., Nishijima, Y.: Bull. Chem. Soc. Jpn. *54*, 35 (1981)
13. De Schryver, F. C., Demeyer, K., van der Auweraer, M., Quanten, E.: Ann. N. Y. Acad. Sci. *366*, 93 (1981)
14. Smakula, A.: Angew. Chem. *47*, 777 (1934)
15. Vala, M. T., Jr., Rice, S. A.: J. Chem. Phys. *39*, 2348 (1963)
16. Hirayama, F., Basile, L. J., Kikuchi, C.: Mol. Cryst. *4*, 83 (1968)
17. Gargallo, L., Abuin, E. A., Lissi, E. A.: Scientia (Valparaiso) *42*, 11 (1977)
18. Longworth, J. W.: Biopolymers *4*, 1131 (1966)
19. Gény, F., Noel, C., Monnerie, L.: J. Chim. Phys. *71*, 1150 (1974)
20. Yanari, S. S., Bovey, F. A., Lumry, R.: Nature *200*, 242 (1963)
21. McKellar, J. F., Allen, N. S.: "Photochemistry of Man-Made Polymers", Applied Science Publ.: London 1979, pp. 23–25
22. Klöpffer, W.: Eur. Polym. J. *11*, 203 (1975)
23. Reference 3, Chapter 10
24. Chien, J. C. W.: J. Phys. Chem. *69*, 4317 (1965)
25. Lindsell, W. E., Robertson, F. C., Soutar, I.: Eur. Polym. J. *17*, 203 (1981)

26. Laitinen, H. A., Miller, F. A., Parks, T. D.: J. Am. Chem. Soc. *69*, 2707 (1947)
27. Nishijima, Y., Yamamoto, M., Mitani, K., Katayama, S., Tanibuchi, T.: Rep. Prog. Polym. Phys. Jpn. *13*, 417 (1970)
28. Nishijima, Y., Yamamoto, M., Katayama, S., Hirota, K., Sasaki, Y., Tsujisaki, M.: ibid. *15*, 445 (1972)
29. Klemm, L. H., Kohlik, A. J.: J. Org. Chem. *28*, 2044 (1963)
30. Brenner, S., Bovere, M.: Tetrahedron *31*, 153 (1975)
31. Fox, R. B., Price, T. R., Cozzens, R. F., McDonald, J. R.: J. Chem. Phys. *57*, 534 (1972)
32. David, C., Baeyens-Volant, D., Geuskens, G.: Eur. Polym. J. *12*, 71 (1976)
33. Irie, M., Kamijo, T., Aikawa, M., Takemura, T., Hayashi, K., Baba, H.: J. Phys. Chem. *81*, 1571 (1977)
34. Pratte, J. F., Webber, S. E.: Macromolecules *15*, 417 (1982)
35. Beale, R. N., Roe, E. M. F.: J. Chem. Soc. 2884 (1951)
36. Nowakowska, M., Najbar, J., Waligora, B.: Eur. Polym. J. *12*, 387 (1976)
37. Klöpffer, W., in: "Organic Molecular Photophysics"; Birks, J. B. (Ed.) Wiley-Interscience: New York 1973, Vol. 1, Chapter 7
38. Zachariasse, K. A., Kühnle, W.: Z. Phys. Chem. N. F. *101*, 267 (1976)
39. Davidson, R. S., Whelan, T. D.: J. Chem. Soc., Chem. Commun. 361 (1977)
40. Ito, S., Yamamoto, M., Nishijima, Y.: Rep. Prog. Polym. Phys. Jpn. *22*, 453 (1979)
41. Ito, S., Yamamoto, M., Nishijima, Y.: ibid *22*, 457 (1979)
42. Reference 3, p. 126
43. Abuin, E. A., Lissi, E. A., Gargallo, L., Radic, D.: Eur. Polym. J. *16*, 793 (1980)
44. Phillips, D., Roberts, A. J., Soutar, I.: J. Polym. Sci., Polym. Phys. Ed. *18*, 2401 (1980)
45. Phillips, D., Roberts, A. J., Soutar, I.: Polymer *22*, 293 (1981)
46. Jones, P. F., Calloway, A. R., in: "Transitions Non Radiatives dans les Molécules", 20th Mtg. Soc. Chim. Phys., May 27–30, 1969; Soc. Chim. Phys.: Paris 1970, pp. 110–115
47. Offen, H. W., Phillips, D. T.: J. Chem. Phys. *49*, 3995 (1968)
48. Kim, J. J., Beardslee, R. A., Phillips, D. T.,Offen, H. W.: ibid. *51*, 2761 (1969)
49. Reference 3, p. 508
50. Cozzens, R. F., Fox, R. B.: Polym. Prepr. Am. Chem. Soc., Div. Polym. Chem. *9(1)*, 363 (1968)
51. Fox, R. B., Price, T. R., Cozzens, R. F.: J. Chem. Phys. *54*, 79 (1971)
52. Fox, R. B., Price, T. R., Cozzens, R. F., McDonald, J. R.: ibid. *57*, 2284 (1972)
53. David, C., Lempereur, M., Geuskens, G.: Eur. Polym. J. *8*, 417 (1972)
54. Burkhart, R. D., Aviles, R. G., Magrini, K.: Macromolecules *14*, 91 (1981)
55. Fox, R. B., Price, T. R., Cozzens, R. F., Echols, W. E.: ibid. *7*, 937 (1974)
56. Pasch, N. F., Webber, S. E.: Chem. Phys. *16*, 361 (1976)
57. Kim, N., Webber, S. E.: Macromolecules *13*, 1233 (1980)
58. Ito, S., Nishino, S., Yamamoto, M., Nishijima, Y.: Rep. Prog. Polym. Phys. Jpn. *24*, 481 (1981)
59. Fox, R. B., Price, T. R., Cozzens, R. F., McDonald, J. R.: J. Chem. Phys. *57*, 534 (1972)
60. Pratte, J. F., Webber, S. E.: Polym. Prepr., Am. Chem. Soc., Div. Polym. Chem. *20(1)*, 927 (1979)
61. Bensasson, R. V., Ronfard-Haret, J. C., Land, E. J., Webber, S. E.: Chem. Phys. Lett. *68*, 438 (1979)
62. Jones, P. F., Siegel, S.: J. Chem. Phys. *50*, 1134 (1969)
63. Czarnecki, S.: Bull. Acad. Polon. Sci., Ser. Math. Astron. Phys. *9*, 561 (1961)
64. Oster, G., Geacintov, N., Khan, A. U.: Nature *196*, 1089 (1962)
65. Reference 3, Chapter 6
66. Klöpffer, W.: J. Chem. Phys. *50*, 2337 (1969)
67. Jones, P. F.: J. Polym. Sci., Polym. Lett. Ed. *6*, 487 (1968)
68. Birks, J. B.: Prog. Reaction Kinetics *5*, 181 (1970)
69. Birks, J. B.: Rep. Prog. Phys. *38(3)*, 903 (1975)
70. Schweitzer, D., Colpa, J. P., Behnke, J., Hausser, K. H., Haenel, M., Staab, H. A.: Chem. Phys. *11*, 373 (1975)
71. Reference 3, Chapter 7
72. Malar, E. J. P., Chandra, A. K.: Theor. Chim. Acta *55*, 153 (1980)
73. Reference 3, Chapter 9

74. Braun, H., Förster, Th.: Ber. Buns. Phys. Chem. *70*, 1091 (1966)
75. Selinger, B. K.: Aust. J. Chem. *19*, 825 (1966)
76. Cundall, R. B., Pereira, L. C.: Chem. Phys. Lett. *15*, 383 (1972)
77. Aladekomo, J. B., Birks, J. B.: Proc. Royal Soc. pt. A *284*, 551 (1965)
78. Stevens, B.: Spectrochim. Acta *18*, 439 (1962)
79. Robertson, J. M.: "Organic Crystals and Molecules", Cornell University Press: Ithaca, New York 1953
80. Ferguson, J.: J. Chem. Phys. *43*, 306 (1965)
81. Jones, P. F., Siegel, S.: ibid. *54*, 3360 (1971)
82. Avis, P., Porter, G.: J. Chem. Soc., Faraday Trans. II *70*, 1057 (1974)
83. Johnson, G. E.: Macromolecules *13*, 839 (1980)
84. Birks, J. B., Kazzaz, A. A., King, T. A.: Proc. Royal Soc. pt. A *291*, 556 (1966)
85. Siegel, S., Stewart, T.: J. Chem. Phys. *55*, 1775 (1971)
86. Kawakubo, T.: J. Phys. Soc. Jpn. *31*, 1446 (1971)
87. Kitaigorodskii, A. I.: "Organic Chemical Crystallography"; Consultants Bureau: New York, pp. 225–227 (1961)
88. Farid, S., Martic, P. A., Daly, R. C., Thompson, D. R., Specht, D. P., Hartman, S. E., Williams, J. L. R.: Pure Appl. Chem. *51*, 241 (1979)
89a. Melzer, G., Schweitzer, D., Hausser, K. H., Colpa, J. P., Haenel, W. M.: Chem. Phys. *39*, 229 (1979)
89b. Baker, W. et al.: J. Chem. Soc. 2991 (1952)
90. Cram, D. J.; Dalton, C. K., Knox, G. R.: J. Am. Chem. Soc. *85*, 1088 (1963)
91. Wasserman, H. H., Keehn, P. M.: J. Am. Chem. Soc. *91*, 2374 (1969)
92. Froines, J. R., Hagerman, P. J.: Chem. Phys. Lett. *4*, 135 (1969)
93. Kaupp, G., Zimmermann, I.: Angew. Chem. Int. Ed. Engl. *15*, 441 (1976)
94. Ishikawa, S.: Bull. Chem. Soc. Jpn. *52*, 1346 (1979)
95. Haenel, M. W., Staab, H. A.: Tetrahedron Lett. 3585 (1970)
96. Haenel, M. W., Staab, H. A.: Chem. Ber. *106*, 2203 (1973)
97. Blank, N. E.: Ph. D. thesis, University of Heidelberg, 1980
98. Haenel, M. W.: Tetrahedron Lett. 3053 (1974)
99. Schweitzer, D., Colpa, J. P., Hausser, K. H., Haenel, M. W., Staab, H. A.: J. Luminescence *12/13*, 363 (1976)
100. Haenel, M. W.: Chem. Ber. *111*, 1789 (1978)
101. Boekelheide, V., Tsai, C. H.: Tetrahedron *32*, 423 (1976)
102. Iskander, M. N., Reiss, J. A.: Tetrahedron *34*, 2343 (1978)
103. Blank, N. E., Haenel, M. W.: Tetrahedron Lett. 1425 (1978)
104. Blank, N. E., Haenel, M. W.:Chem. Ber. *114*, 1520 (1981)
105. Kawabata, T., Shinmyozu, T., Inazu, T., Yoshino, T.: Chem. Lett. 315 (1979)
106. Yoshinaga, M., Otsubo, T., Sakata, Y., Misumi, S.: Bull. Chem. Soc. Jpn. *52*, 3759 (1979)
107. Nishimura, J., Furukawa, M., Yamashita, S., Inazu, T., Yoshino, T.: J. Polym. Sci., Polym. Chem. Ed. *19*, 3257 (1981)
108. Blank, N. E., Haenel, M. W.: Chem. Ber. *114*, 1531 (1981)
109. Chandross, E. A.: J. Chem. Phys. *43*, 4175 (1965)
110. Ferguson, J., Mau, A. W. H., Morris, J. M.: Aust. J. Chem. *26*, 91 (1973)
111. Chandross, E. A., Dempster, C. J.: J. Am. Chem. Soc. *92*, 704 (1970)
112. Todesco, R., Gelan, J., Martens, H., Put, J., Boens, N., De Schryver, F. C.: Tetrahedron Lett. *31*, 2815 (1978)
113. Todesco, R., Gelan, J., Martens, H., Put, J., De Schryver, F. C.: Bull. Soc. Chim. Belg. *89*, 521 (1980)
114. Todesco, R., Gelan, J., Martens, H., Put, J., De Schryver, F. C.: J. Am. Chem. Soc. *103*, 7304 (1981)
115. Goldenberg, M., Emert, J., Morawetz, H.: ibid. *100*, 7171 (1978)
116. Itagaki, H., Obukata, N., Okamoto, A., Horie, K., Mita, I.: Chem. Phys. Lett. *78*, 143 (1981)
117. Kamijo, T., Irie, M., Hayashi, K.: Bull. Chem. Soc. Jpn. *51*, 3286 (1978)
118. Ibemesi, J. A., Kinsinger, J. B., El-Bayoumi, M. A.: J. Polym. Sci., Polym. Chem. Ed. *18*, 879 (1980)
119. Zachariasse, K. A., Kühnle, W.: Z. Phys. Chem. N. F. *101*, 267 (1976)

120. Redpath, A. E. C., Winnik, M. A.: Ann. N. Y. Acad. Sci. *366*, 75 (1981)
121. Zachariasse, K. A., Kühnle, W., Weller, A.: Chem. Phys. Lett. *59*, 375 (1978)
122. Braun, H., Förster, Th.: Z. Phys. Chem. N. F. *78*, 40 (1972)
123. Wang, Y.-C., Morawetz, H.: J. Am. Chem. Soc. *98*, 3611 (1976)
124. Bokobza, L., Jasse, B., Monnerie, L.: Eur. Polym. J. *16*, 715 (1980)
125. Emert, J., Behrens, C., Goldenberg, M.: J. Am. Chem. Soc. *101*, 771 (1979)
126. Flory, P. J.: "Statistical Mechanics of Chain Molecules", Wiley-Interscience: New York 1969
127. Fitzgibbon, P. D., Frank, C. W.: Macromolecules *14*, 1650 (1981); Fitzgibbon, P. D. Ph. D. Thesis, Stanford University 1980
128. Nakahira, T., Ishizuka, S., Iwabuchi, S., Kojima, K.: Macromolecules *15*, 1217 (1982)
129. Avouris, P., Kordas, J., El-Bayoumi, M. A.: Chem. Phys. Lett. *26*, 373 (1974)
130. Gupta, A., Liang, R., Moacanin, J., Kliger, D., Goldbeck, R., Horwitz, J., Miskowski, V. M.: Eur. Polym. J. *17*, 485 (1981)
131. Gleria, M., Barigelletti, F., Dellonte, S., Lora, S., Minto, F., Bortolus, P.: Chem. Phys. Lett. *83*, 559 (1981)
132. Frank, C. W., Harrah, L. A.: J. Chem. Phys. *61*, 1526 (1974)
133. De Schryver, F. C., Vandendriessche, J., Toppet, S., Demeyer, K., Boens, N.: Macromolecules *15*, 406 (1982)
134. Longworth, J. W., Bovey, F. A.: Biopolymers *4*, 1115 (1966)
135. De Schryver, F. C., Moens, L., Van der Auweraer, M., Boens, N., Monnerie, L., Bokobza, L.: Macromolecules *15*, 64 (1982)
136. Stegen, G. E., Boyd, R. E.: Polym. Prepr., Am. Chem. Soc., Div. Polym. Chem. *19(1)*, 595 (1978) and references within
137. Ito, S., Yamamoto, M., Nishijima, Y.: Bull. Chem. Soc. Jpn. *55*, 363 (1982)
138. Seki, K., Ichimura, Y., Imamura, Y.: Macromolecules *14*, 1831 (1981)
139. Seki, K., Ichimura, Y., Imamura, Y.: Rep. Prog. Polym. Phys. Jpn. *23*, 595 (1980)
140. Gelles, R., Frank, C. W.: Macromolecules *15*, 741 (1982)
141. Fitzgibbon, P. D., Frank, C. W.: ibid *15*, 733 (1982)
142. Yoon, D. Y., Sundararajan, P. R.; Flory, P. J.: ibid. *8*, 776 (1975)
143. Richards, D. H., Scilly, N. S., Williams, F.: Polymer *10*, 603 (1969)
144. David, C., Lempereur, M., Geuskens, G.: Eur. Polym. J. *9*, 1315 (1973)
145. David, C., Lempereur, M., Geuskens, G.: ibid. *10*, 1181 (1974)
146. Yokoyama, M., Tamamura, T., Atsumi, M., Yoshimura, M., Shirota, Y., Mikawa, H.: Macromolecules *8*, 101 (1975)
147. Jachowicz, J., Morawetz, H.: ibid. *15*, 828 (1982)
148. Gelles, R., Frank, C. W.: ibid. *15*, 747 (1982)
149. Somersall, A. C., Guillet, J. E.: ibid. *6*, 218 (1973)
150. Aspler, J. S., Guillet, J. E.: ibid. *12*, 1082 (1979)
151. Holden, D. A., Wang, P. Y.-K., Guillet, J. E.: ibid. *13*, 295 (1980)
152. Merle-Aubry, L., Holden, D. A., Merle, Y., Guillet, J. E.: ibid. *13*, 1138 (1980)
153. Abuin, E. A., Lissi, E. A., Gargallo, L., Radic, D.: Eur. Polym. J. *15*, 373 (1979)
154. Nakahira, T., Maruyama, I., Iwabuchi, S., Kojima, K.: Makromol. Chem. *180*, 1853 (1979)
155. Nakahira, T., Ishizuka, S., Iwabuchi, S., Kojima, K.: Makromol. Chem., Rapid Commun. *1*, 437 (1980)
156. Lentz, P., Blume, H., Schulte-Frohlinde, D.: Ber. Buns. Phys. Chem. *74*, 484 (1970)
157. Webber, S. E., Avots-Avotins, P. E., Deumie, M.: Macromolecules *14*, 105 (1981)
158. Ito, S., Yamamoto, M., Nishijima, Y.: Polymer J. *13*, 791 (1981)
159. Zachariasse, K. A.: private communication
160. Ghiggino, K. P., Wright, R. D., Phillips, D.: J. Polym. Sci., Polym. Phys. Ed. *16*, 1499 (1978)
161. MacCallum, J. R., Rudkin, L.: Eur. Polym. J. *17*, 953 (1981)
162. Kryszewski, M., Wandelt, B., Birch, D. J. S., Imhof, R. E., North, A. M., Pethrick, R. A.: Polymer *23*, 924 (1982)
163. Yamamoto, M., Hirota, K., Nishijima, Y.: Rep. Prog. Polym. Phys. Jpn. *13*, 429 (1970)
164. Ito, S., Yamamoto, M., Nishijima, Y.: ibid. *20*, 481 (1977)
165. Ito, S., Yamamoto, M., Nishijima, Y.: ibid. *21*, 393 (1978)
166. Ito, S., Yamamoto, M., Nishijima, Y.: ibid. *19*, 421 (1976)

167. Johnson, G. E., Good, T. A.: Macromolecules *15*, 409 (1982)
168. Koski, W. S.: Phys. Rev. *82*, 230 (1951)
169. Reference 3, Chapter 11; and references therein
170. Klöpffer, W.: Spectr. Lett. *11*, 863 (1978)
171. Klöpffer, W.: Ann. N. Y. Acad. Sci. *366*, 373 (1981)
172. Turro, N. J.: Pure Appl. Chem. *49*, 405 (1977)
173. MacCallum, J. R.: Annu. Rep. Prog. Chem., Sect. A: Phys. Inorg. Chem. *75*, 99 (1978)
174. Nishijima, Y., Mitani, K., Katayama, S., Yamamoto, M.: Rep. Prog. Polym. Phys. Jpn. *13*, 421 (1970)
175. Ishii, T., Handa, T., Matsunaga, S.: Macromolecules *11*, 40 (1978)
176. Torkelson, J. M., Lipsky, S., Tirrell, M.: ibid. *14*, 1603 (1981)
177. Förster, Th.: Comprehensive Biochem. *22*, 61 (1967)
178. David, C., Lempereur, M., Geuskens, G.: Eur. Polym. J. *9*, 1315 (1973)
179. Reid, R. F., Soutar, I.: J. Polym. Sci., Polym. Phys. Ed. *16*, 231 (1978)
180. Hill, D. J. T., Lewis, D. A., O'Donnell, J. H., O'Sullivan, P. W., Pomery, P. J.: Eur. Polym. J. *18*, 75 (1982)
181. Nishijima, Y., Mitani, K., Katayama, S., Yamamoto, M.: Rep. Prog. Polym. Phys. Jpn. *13*, 425 (1970)
182. David, C., Piens, M., Geuskens, G.: Eur. Polym. J. *12*, 621 (1976)
183. Anderson, R. A., Reid, R. F., Soutar, I.: ibid. *15*, 925 (1979)
184. Alexandru, L., Somersall, A. C.: J. Polym. Sci., Polym. Chem. Ed. *15*, 2013 (1977)
185. McInally, I., Reid, R. F., Rutherford, H., Soutar, I.: Eur. Polym. J. *15*, 723 (1979)
186. MacCallum, J. R.: ibid. *17*, 209 (1981)
187. David, C., Putman, N., Geuskens, G.: ibid. *13*, 15 (1977)
188. MacCallum, J. R., Rudkin, L.: Nature *266*, 338 (1977)
189. MacCallum, J. R.: Polymer *23*, 175 (1982)
190. Heisel, F., Laustriat, G.: J. Chim. Phys. *66*, 1881 (1969)
191. Ishii, T., Handa, T., Matsunaga, S.: Makromol. Chem. *178*, 2351 (1977)
192. Ishii, T., Utena, Y., Handa, T., Mori, S., Takagi, K.: Rep. Prog. Polym. Phys. Jpn. *20*, 415 (1977)
193. Ishii, T., Handa, T., Mori, S., Utena, Y.: ibid. *21*, 361 (1978)
194. Ito, S., Shiga, T., Yamamoto, M., Nishijima, Y.: ibid. *23*, 543 (1980)
195. Phillips, D., Roberts, A. J., Soutar, I.: Polymer *22*, 427 (1981)
196. Keyanpour-Rad, M., Ledwith, A., Hallam, A., North, A. M., Breton, M., Hoyle, C. E., Guillet, J. E.: Macromolecules *11*, 1114 (1978)
197. Ledwith, A., Rowley, N. J., Walker, S. M.: Polymer *22*, 435 (1981)
198. Ng, D., Guillet, J. E.: Macromolecules *15*, 724 and 728 (1982)
199. Abuin, E. A., Lissi, E. A., Gargallo, L., Radic, D.: Eur. Polym. J. *15*, 373 (1979)
200. Bowen, E. J., Brocklehurst, B.: Trans. Faraday Soc. *49*, 1131 (1953)
201. Bowen, E. J., Livingston, R.: J. Am. Chem. Soc. *76*, 6300 (1954)
202. Bowen, E. J., Brocklehurst, B.: Trans. Faraday Soc. *51*, 774 (1955)
203. Berlman, I. B.: "Energy Transfer Parameters of Aromatic Compounds", Academic Press: New York 1973
204. Förster, Th.: Disc. Faraday Soc. *27*, 7 (1959); and references therein
205. Dale, R., Eisinger, J., in: "Biochemical Fluorescence: Concepts", Vol. 1; Chen, R. F., Edelhoch, H. (Eds.) Marcel Dekker: New York 1975, p. 115
206. Hoyle, C. E., Guillet, J. E.: J. Polym. Sci., Polym. Lett. Ed. *16*, 185 (1978)
207. Aspler, J. S., Hoyle, C. E., Guillet, J. E.: Macromolecules *11*, 925 (1978)
208. Guillet, J. E.: Polym. Prepr., Am. Chem. Soc., Div. Polym. Chem. *20(1)*, 395 (1979)
209. Holden, D. A., Guillet, J. E.: Macromolecules *13*, 289 (1980)
210. Thomas, J. W., Jr., Frank, C. W., Holden, D. A., Guillet, J. E.: J. Polym. Sci., Polym. Phys. Ed. *20*, 1749 (1982)
211. Hargreaves, J. S., Webber, S. E.: Macromolecules *15*, 424 (1982)
212. Powell, R. C., Soos, Z. G.: J. Luminescence *11*, 1 (1975)
213. Fredrickson, G. H., Frank, C. W.: Macromolecules *16*, 1198 (1983)

Received May 11, 1983
J. D. Ferry (editor)

Fourier Transform Infrared Spectroscopy of Polymers

Jack L. Koenig
Department of Macromolecular Science, Case Western Reserve University, Cleveland, Ohio 44106, U.S.A.

This review covers the theory and application of Fourier transform infrared spectroscopy to the characterization of polymers. The basic theory, the sampling techniques and the spectral operations are described. The applications discussed include the study of polymer reactions, polymer structure and dynamic effects.

Advances in Polymer Science 54

1 Introduction

Infrared spectroscopy has long been recognized as a powerful tool for the characterization of polymers [1]. The review by Krimm [2] clearly documented the uses of dispersion infrared spectroscopy for polymer studies. The advent of Fourier transform infrared spectroscopy (FT-IR) has brought about a revival of interest in infrared spectroscopy as a characterization technique. The increased speed and higher signal-to-noise ratio of FT-IR relative to dispersion infrared has lead to a substantially greater number of applications of infrared in polymer research. Also the availability of a dedicated computer, which is required for the FT-IR instrumentation, has allowed the digitized spectra to be treated by sophisticated data processing techniques and has increased the utility of the infrared spectra for qualitative and quantitative purposes. So with interferometric techniques infrared spectroscopy is being launched into a new era and interest in the technique is at an all time high.

FT-IR has almost become a ubiquitous technique in physical and analytical chemistry and a number of sources can be consulted for details [3,4,5,6]. This review will only consider those applications related to polymer systems. A number of prior reviews of the applications of FT-IR have been written [7,8,9,10,11,12,13,14]. The relevant theory of vibrational spectroscopy of polymers will not be covered here as it has been treated in detail in a recent monograph [15]. An excellent discussion of the experimental aspects of the spectroscopy of polymers is also recommended [16].

2 The FT-IR Method

Infrared spectroscopy is an old and familiar technique for polymer characterization. It is based on the absorption of radiation in the infrared frequency range due to

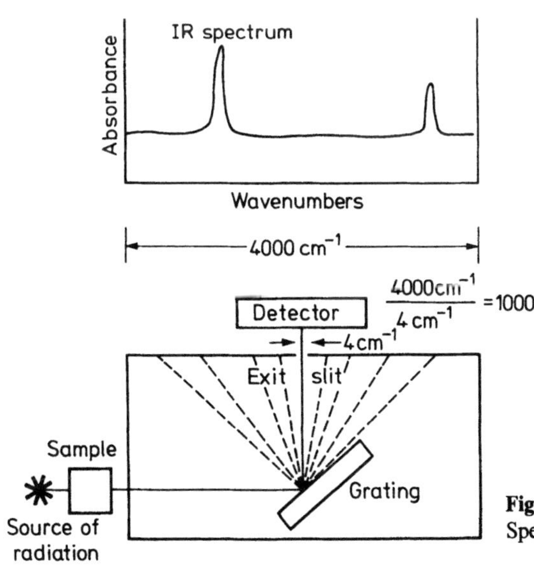

Fig. 1. Optical Diagram of the IR Dispersive Spectrometer

the molecular vibrations of the functional groups contained in the polymer chain. Prior to FT-IR, infrared spectroscopy was carried out using a dispersive instrument utilizing prisms or gratings to geometrically disperse the infrared radiation (Fig. 1). Using a scanning mechanism, the dispersed radiation was passed over a slit system which isolated the frequency range falling on the detector. In this manner, the spectrum, that is, the energy transmitted through a sample as a function of frequency, was obtained. This infrared method is highly limited in sensitivity because most of the available energy is being thrown away, i.e. does not fall on the open slits. Hence, to improve the sensitivity of infrared spectroscopy, a technique is sought which allows the examination of all of the transmitted energy all of the time.

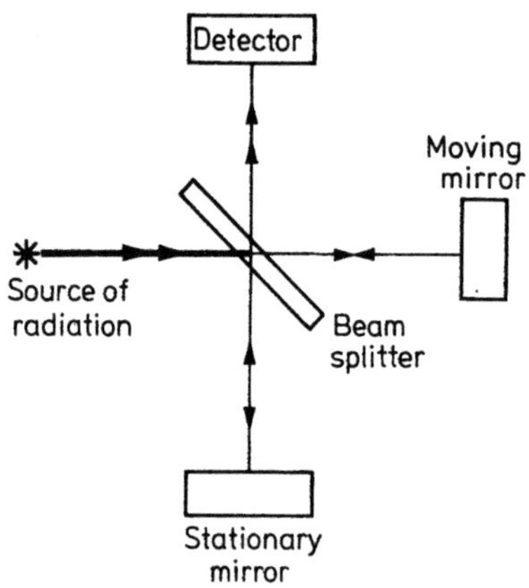

Fig.2. Optical Diagram of the Interferometer

The Michelson interferometer is such an optical device[17,18], (Fig. 2). Several excellent sources are available for reference to instrumentation theory and practive with interferometers[19,20,21,22,23]. Only a brief description will be given here. The Michelson interferometer has two mutually perpendicular arms. One arm of the interferometer contains a stationary, plane mirror; the other arm contains a movable mirror. Bisecting the two arms is a beamsplitter which splits the source beam into two equal beams. These two light beams travel down their respective arms of the interferometer and are reflected back to the beamsplitter and on to the detector. The two reunited beams will interfere constructively or destructively, depending on the relationship between their path difference (x) and the wavelengths of light. When the movable mirror and the stationary mirror are positioned the same distance from the beamsplitter in their respective arms of the interferometer ($x = 0$), the paths of the light beams are identical. Under these conditions all wavelengths of the radiation striking the beam splitter after reflection add coherently to produce a maximum flux at the detector and generate what is known as the "center burst". As the movable mirror is displaced from this point, the path length in that arm of the interferometer is changed. This difference in path length causes each wavelength of source radiation to destructively interfere with itself at the beam splitter. The resulting flux at the detector, which is the sum of the fluxes for each of the individual wavelengths, rapidly decreases with mirror displacement. By sampling the flux at the detector, one obtains an interferogram. For a monochromatic source of frequency v, the interferogram is given by the expression

$$I(x) = 2RTI(v)\,(1 + \cos 2\pi\, vx)$$

where R is the reflectance of the beam splitter, T is the transmittance of the beam splitter, I(v) is the input energy at frequency v and x is the path difference. The interferogram consists of two parts, a constant (DC) component equal to 2RTI, and a modulated (AC) component. The AC component is called the the interferogram and is given by

$$I(x) = 2RTI(v)\cos(2\pi\, vx)$$

An infrared detector and AC-amplifer converts this flux into an electrical signal

$$V(x) = re\, I(x) \qquad \text{volts}$$

where re is the response of the detector and amplifier.

The moving mirror is generally driven on an air bearing and for FT-IR measurements the mirror must be kept in the same plane to better than 10 micro-radians for mirror drives up to 10 cm^{-1} in length. It is necessary to have some type of marker to initiate data acquisition at precisely the same mirror displacement everytime. The uncertainty in this position cannot be greater than 0.1 microns from scan to scan. This precise positioning is accomplished by having on the moving mirror a smaller reference interferometer. Through this reference interferometer is passed a visible white light source. This white light source has a very sharp "center burst" or spike and the data acquisition is initiated when this signal reaches a predetermined value.

Digitization of the interferogram at precisely spaced intervals is accomplished with the help of the interference signal of the auxiliary interferometer equipped with a He-Ne laser which yields a monochromatic signal resulting in the output of a cosine function from the interferometer (Fig. 3). The interferogram is sampled at each zero-crossing of the laser cosine function. Therefore, the path difference between two successive data points in the digitized interferogram is always a multiple of half wavelengths of the laser, or 0.316 microns. The laser also provides an internal calibration of the wavelength.

For highest accuracy in the digitized signal, the maximum intensity in an interferogram should match as closely as possible to the maximum input voltage of the analog/digital converter (ADC). The noise must also be given at least four or five units, so the computer word length in FT-IR spectrometers is 16 (or 32 in double precision), 20 and 24 bits. Griffiths [19)] gives the example of measuring the spectrum of a continuous source whose intensity is uniform from 4000 to 400 cm^{-1} and zero outside this band. The signal to noise ratio of the spectrum $(S/N)_S$ is related to the signal-to-noise ratio of the interferogram $(S/N)_I$ by

$$(S/N)_I = M^{1/2} \cdot (S/N)_S$$

where M is the number of resolution elements. So if we want to measure the spectrum with a $(S/N)_S = 500$ at resolution 1 cm^{-1} ($M = 3600$) in a single scan, it can readily

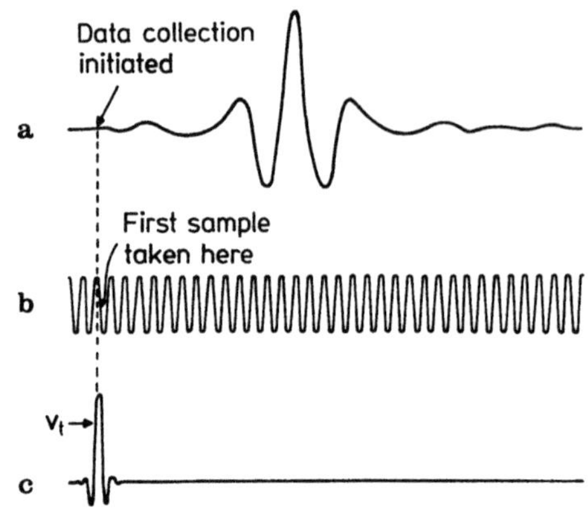

Fig. 3a—c. Signals from an interferometer with a separate reference cube for measuring the laser and white light interferograms. **a)** Signal (infrared) interferogram; **b)** laser interferogram; **c)** white light interferogram. The zero retardation position of the reference interferometer has been displayed relative to that of the main interferometer so that data collection may be initiated a short distance before the maximum of the infrared interferogram.
Ref. Griffiths, P. R., "Fourier Transform Infrared Spectrometry: Theory and Instrumentation", in *Transform Techniques in Chemistry*, P.R. Griffiths, ed., Plenum Press, New York (1978) p. 125

be seen that $(S/N)_I = 3 \times 10^4$ and the full dynamic range of a 15-bit ADC (2^{15}) is only large enough to adequately digitize the signal. If $(S/N)_I$ were any greater than 30,000 the "noise level" in the digitized interferogram is set by the least significant bit of the ADC rather than by detector noise. For this reason, minicomputers with a wordlength as large as 20 bits are required so that a reasonable number of scans can be signal averaged.

The interferogram for a polychromatic source A(v) is given by

$$I(x) = \int_0^\infty A(v)\,(1 + \cos 2\pi v x)\,dv$$

The methods of evaluating these integrals involve a determination of the values at zero-path length and very long or infinite path length. At zero difference

$$\overset{0}{I(x)} = 2\int_0^\infty A(v)\,dv$$

and for large path differences

$$\overset{\infty}{I(x)} = \int_0^\infty A(v)\,dv = I(0)/2$$

so the actual interferogram F(x) is

$$F(x) = I(x) - I(\infty) = \int_0^\infty A(v)\cos(2\pi v x)\,dv\,.$$

From Fourier transform theory [24,25,26,27,28],

$$A(v) = 2 \int_{-0}^{+\infty} F(x) \cos(2\pi v x)\, dx$$

This Fourier transform process was well known to Michelson and his peers but the computational difficulty of making the transformation prevented the application of this powerful interferometric technique to spectroscopy. An important advance was made with the discovery of the fast Fourier transform algorithm by Cooley and Tukey [29] which revived the field of spectroscopy using interferometers by allowing the calculation of the Fourier transform to be carried out rapidly. The fast Fourier transform (FFT) has been discussed in several places [30,31]. The essence of the technique is the reduction in the number of computer multiplications and additions. The normal computer evaluation requires $n(n-1)$ additions and multiplications whereas the FFT method only requires $(n \log_2 n)$ additions and multiplications. If we have a 4096-point array to Fourier transform, it would require (4096) (4095) or 16.7 million multiplications. The FFT allows us to reduce this to

$$(4096) \cdot \log_2 (4096) \text{ or } 4096(12) = 49{,}153 \text{ multiplications},$$

a saving by a factor of 341 in time [32]. The advantage of the FFT increases with the number of data points. As computers have improved, the time required for a Fourier transform has reduced until currently the transformation can be carried out in less than a second with fast array processors. Thus the spectra can be calculated during the retrieval of the moving mirror if desired.

However, it is to be noted that the Fourier transform integrals have infinite limits while the optical path differences are finite so modifications or approximations must be made. We will use the approximation of the limits between L and +L where L is the maximum distance of the mirror drive. So

$$S(v) = 2 \int_{-L}^{+L} F(x) \cos(2\pi v x)\, dx$$

where S(v) is used to indicate that we are approximating the Fourier transform. It is of interest to examine the effect of this finite optical path length approximation on the $S(v_k)$ of an incident monochromatic source of wavelength v_k. The interferogram for this source is

$$F(x) = A(v_k) \cos 2\pi\, v_k x$$

where $A(v_k)$ is the amplitude of the light intensity. Making the substitution, we obtain

$$S(v_k) = 2 \int_{-L}^{+L} A(v_k) \cos(2 v_k x) * \cos(2\pi v x)\, dx$$

and after the transformation one obtains

$$S(v_k) = A(v_k) \cdot 2L \operatorname{sinc}^2 \pi(v_k - v)\, L$$

where the sinc y function of y is siny/y. This function is shown in Fig. 4a and represents the instrument line shape of a Michelson interferometer. Obviously, this instrument line shape is not satisfactory because of the strong side lobes. These side lobes can be removed by apodization [33]. Apodization is carried out by multiplying the interferogram by a function H(x), before the Fourier interferogram is transformed. Thus

$$S(v)\,2 = \int_{-L}^{+L} F(x) * H(x) \cos(2v\pi x)\,dx\,.$$

Although a variety of apodization functions have been examined [34,35,36], triangular apodization has the following form

$$H(x) = 1 - |(x/L)| \text{ if } |x| \leqq L \text{ and zero otherwise}\,.$$

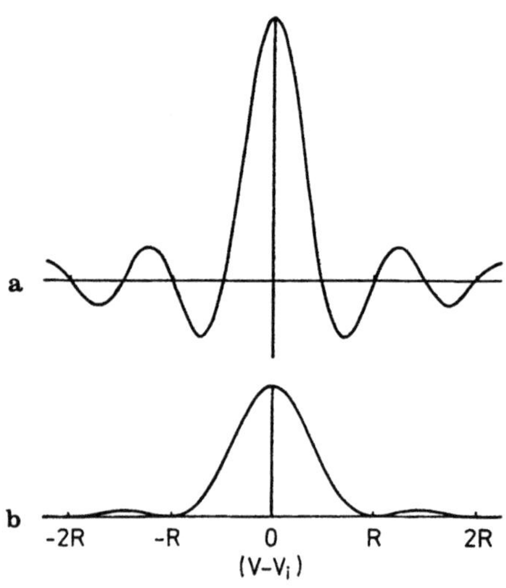

Fig. 4. The functions **a)** $I(V_k)$ and **b)** $S(V_k)$, which are the instrument line shape functions for spectra computed using no apodization and triangular apodization, respectively.
Ref. Griffiths, P. R., "Fourier Transform Infrared Spectrometry: Theory and Instrumentation", in *Transform Techniques in Chemistry*, P. R. Griffiths, ed., Plenum Press, New York (1978) p.114

Again for a monochromatic source, we have

$$S(v) = 2\int_{-\infty}^{+\infty} A(v_k)\,(1 - x/L)\cos(2\pi v_k x) * \cos(2\pi v_k x)\,dx\,.$$

Note that the integration limits could now be changed to plus and minus infinity without changing the result, because the integrand f is zero outside of the range $-L \leqq x \leqq L$ integration [37] yields

$$S(v) = A(v_k) \cdot L \cdot \mathrm{sinc}^2\,[\pi(v_k - v)\,L]$$

and is shown in (Fig. 4b). A summary of the effects of apodization has been given [38] and recommendations made for the appropriate apodization function for quantitative infrared measurements.

There are a number of other problems such as phase correction arising from the fact that in practice, the radiation undergoes phase shifts due to beamsplitter characteristics, signal processing delays, refraction effects in materials and interferometer misalignment [39]. However, techniques have been developed which allow appropriate corrections to be made for these effects. Mertz [40] has developed a method for use in the spectral domain and Forman et al. [41] a method to be performed on the interferogram. Recently a rapid practical method of making the phase correction has been described [42] using a Fourier convolution method.

The typical FT-IR instrument has a number of components including a minicomputer, a moving head disc, ADC capable of operating up to 50 kHz, input-output interfaces, data terminal for input, a high speed digital plotter, and an oscilloscope with graphic and alphanumeric display. For software, all commercial systems have complete data processing capabilities including signal averaging, apodization, phase correction, gain-ranging and a variety of programs to manipulate and display the spectral data. With every month, new software programs are being developed and exchanged between the users of the equipment. A number of companies offer turn-key instruments with a wide range of resolution, sensitivity, frequency range and speed. With the large commercial market for FT-IR, continued improvement in hardware and software can be expected.

3 Comparison of FT-IR with Dispersive Infrared Spectroscopy

The advantages of FT-IR over dispersive (i.e. grating) infared arises from several sources. Fellgett's advantage [43] or the multiplex advantage is the principal advantage of FT-IR. For measurements taken at equal resolution and for equal measurement time with the same detector and optical throughput, the signal-to-noise (S/N) of spectra from an FT-IR will be $M^{1/2}$ times greater than on a grating instrument where M is the number of resolution elements being examined during the measurement. Alternately, for a given observation time, it is possible to repeat the FT-IR measurement M times which increases the signal by a factor of M and the noise by a factor of $M^{1/2}$, to achieve a S/N enhancement of a factor of $M^{1/2}$. This advantage arises from the fact that the FT-IR spectrometer examines the entire spectrum in the same period of time required for a dispersive instrument to examine a single spectral element. Theoretically, an FT-IR instrument can acquire the spectrum with the same S/N from 0 to 4000 cm^{-1} with 1 cm^{-1} resolution 4000 times faster than a dispersive instrument. Or from another point of view, for the same measurement time a factor of approximately 63 increase in S/N can be achieved on the FT-IR instrument. Therefore, when there is a limited time for measurement, there is a definite time advantage for the FT-IR instrument. When time of measurement is not an important consideration, the time can be used to multiscan with the FT-IR instrument to signal average and increase the S/N. Of course, there is also the inherent time advantage associated with rapid scanning FT-IR since it requires a very short time to scan the mirror and obtain the complete spectra (ca. 1.5 sec). This time advantage of the FT-IR has been particularly important for the study of polymerization chemistry [44], degradation processes [45] and other time-dependant processes of polymers [46,47] to be discussed later.

The Jacquinot or throughput advantage [48] arises from the loss of energy in the dispersive system due to the gratings and slits. These losses do not occur in an FT-IR instrument which does not contain these elements. Basically, the Jacquinot advantage means that the radiant power of the source is more effectively utilized in interferometers. The throughput for an FT-IR instrument is limited by the size of the mirrors. The Jacquinot's advantage compared for an interferometer and a commercial dispersive spectrometer has been given [49]. This higher throughput is particularly important in the infrared region where the signals are weak since the infrared sources are weak. The throughput advantage has been used for studying strongly absorbing systems such as carbon-black filled rubbers [50] and emission from polymers [51].

The far infared region is difficult to study due to energy limitations. The commonly used blackbody sources, such as Nernst glowers or globars contribute less than one hundredth of one percent of the total energy in the region below 100 cm^{-1} with peak output near 3000 cm^{-1}. An additional problem in the far infrared is the elimination of the unwanted energy from shorter wavelengths. Finally, detectors for the long wavelength region are poor so the overall sensitivity in the far-infrared region is further reduced. But with FT-IR, this region is accessible for study and a number of polymers have been examined in this region as will be discussed later.

The combination of Fellgett's and Jacquinot's advantage coupled with the inherent speed differential should lead to an enormous difference between FT-IR and dispersive instruments. However, in practice, part of this advantage is offset by the difference in the performance of the triglycine sulfate (TGS) and thermocouple detectors. At low modulation frequencies, the thermocouple detector is about an order of magnitude more sensitive than TGS.

The Conne [52] or frequency advantage comes from the fact that the frequencies of an FT-IR instrument are internally calibrated by a laser whereas conventional IR instruments exhibit drifts when changes in alignment occur. This latter advantage is particularly useful for coaddition of spectra to signal average since the frequency accuracy is an absolute requirement in this case. For the absorbance subtraction technique to be useful for samples examined over a period of time such as months or years, frequency accuracy must be maintained. Thus with FT-IR, polymer samples can be scanned, the spectra stored, and years later compared with spectra of the samples run currently. Applications such as quality control and long term aging and weathering immediately come to mind based on the reproducibility of the frequency of an FT-IR instrument over the long term.

Another area of superiority of interferometers is the ease with which stray light is reduced. For a grating monochromator, the stray light arises in two different ways. The largest source occurs when radiation diffracted from one order of a grating reaches the exit slits, radiation reflected from other orders is also present. The other source of stray light arises from reflection in the monochromator of nondispersed light which can reach the detector. There is no way of distinguishing this flux of scattered light from the desired flux so the dynamic range of the detector is decreased. This problem can be severe particularly in energy limited situations. With an interferometer, stray light of the second type which has not been modulated by the interferometer appears at the detector as a constant DC offset, and, as such, has no effect on the computed spectrum. Efficient electronic filtering of the signal which avoids folding effects eliminates this unwanted flux contribution. While there is not direct correspondence to overlapping orders in an interferometer, stray light which has been modulated is also easily handled. If the interferogram is sampled at sufficiently high frequency, energy below this Nyquist frequency will be properly accounted for in the computed spectrum. Since the Fourier components due to high frequency radiation vary most rapidly, it is possible to eliminate them from the interferogram by the inclusion of a simple audio band-pass filter. This type of filtering has the advantage that it occurs after the detector and does not effect the light throughput.

The overall simplicity of an FT-IR compared to a dispersive instrument is also an advantage. For example, a single instrument can be easily converted to study the near, mid or far-infrared frequency region whereas with the dispersive method, three totally different instruments are required. To improve resolution with an FT-IR instrument, the basic design is only slightly modified while for a dispersive instrument different optical components are required.

Of course, FT-IR has inherent disadvantages and perhaps the most important of these is that the raw data, an interferogram, is for all practical purposes unintelligible. Thus a computer is required to interpret the interferogram for us and contact is lost with the "real" data. So the technique approaches the "black box" syndrome. This problem can only be overcome by a thorough understanding of each step in the data collections, mathematical transformation and output. The spectroscopist has to learn computers as well as spectroscopy to effectively operate. Sometimes this additional demand on the spectroscopist is too much and without an in-depth knowledge of the workings of his spectrometer, problems can arise.

4 Data Processing Techniques Using Digitized Infrared Spectra

4.1 Absorbance Subtraction

One of the spectral processing operations most widely used in polymer analysis is the digital subtraction of absorbance spectra in order to reveal or emphasize subtle differences between two samples or a sample and a reference material [7]. The number of polymer applications of absorbance subtraction is rapidly increasing and it is this digital subtraction capability more than any other single factor that has inspired subsequent investigation of polymeric materials using FT-IR. Spectral subtraction is a powerful method of extracting structural information about components of composite spectra [8]. When a polymer is examined before and after a chemical or physical treatment, and subtraction of the original spectrum from the final spectrum is done, positive absorbances reflect the structures that are formed during the treatment and negative absorbances reflect the loss of structure [53]. The advantage of FT-IR difference spectra lies in the ability to compensate for differences in thicknesses of the two samples. This balance of thicknesses allows small spectral differences to be associated with structural changes and not to be outweighed by the differences in the amount of sample in the beams. Additionally, with properly compensated thicknesses, the differences in absorbances can be magnified through computer scale expansion to reveal small details of the spectral differences. The scaling parameter, k, is chosen such that

$$(A_1 - kA_2) = 0$$

where A_1 and A_2 correspond to the absorbances of the internal thickness bands of samples one and two. Multiplication of the spectrum of sample 2 by k will yield a new spectrum having the same optical thickness as sample 1. One may use the peak absorbances, integrated peak areas, or a least-squares curve fitting method to calculate the scaling factor k. The method of choice will depend on the system being examined. It should be remembered that the calculation of the scaling factor cannot be done entirely analytically and the only test of the scaling factor is the resultant difference spectrum, which should be examined carefully before any further analysis is carried out.

Absorbance subtraction can be considered as a spectroscopic separation technique for some problems in polymers. An interesting application in FT-IR difference spectroscopy is the spectral separation of a composite spectrum of a heterophase system. One such example is a semicrystalline polymer which may be viewed as a composite system containing an amorphous and crystalline phase [53]. In general, the infrared spectrum of each of these phases will be different because in the crystalline phase one particular rotational conformation will predominate whereas in the disordered amorphous regions a different rotamer will dominate. Since the infrared spectrum is sensitive to conformations of the backbone, the spectral contributions will be different if they can be isolated. The total absorbance A_t at a frequency v of a semicrystalline polymer may be decomposed into the following contributions

$$A_t(v) = A_c(v) + A_a(v) + A_i(v)$$

where $A_c(v)$ and $A_a(v)$ are the contributions to the total absorbance at frequency v due to the crystalline and amorphous components, respectively, and $A_i(v)$ is the contribution to the total absorbance at frequency v due to phase independent absorptions. For the sample of higher crystallinity

$$A_{1t}(v) = A_{1c}(v) + A_{1a}(v) + A_{1i}(v)$$

while for the lower crystallinity sample one can write a similar equation. Once assignments have been made as to the amorphous and crystalline nature of a band in the spectrum by systematic annealing studies, the two spectra can, with a suitable choice of scaling parameters, be subtracted from each other until bands assigned to one of the phases have been reduced to the background. To generate the pure spectra of the crystalline component one substracts the spectra of the two samples thusl:

$$(A_{1t} - kA_{2t}) = (A_{1c} - kA_{2c}) + (A_{1a} - kA_{2a}) + (A_{1i} - kA_{2i})$$

The scale parameter k is chosen such that

$$(A_{1a} - kA_{2a}) = 0$$

The resultant difference spectrum is the "purified" spectrum of the crystalline phase. A similar set of equations hold for generating the amorphous phase spectrum. This

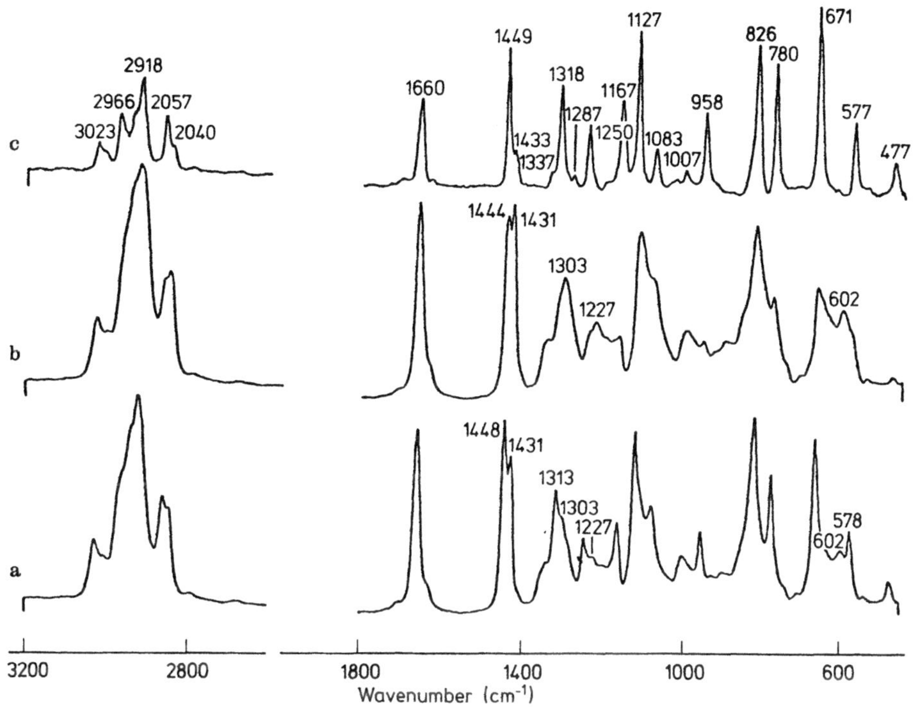

Fig. 5a—c. Infrared Spectra of *Trans*-1,4-Polychloroprenc (—20 °C polymer):
a) Absorbance spectrum at room temperature;
b) Absorbance spectrum at 80 °C;
c) Absorbance spectrum of crystalline vibrational bands of *trans*-1,4-polychloroprene (Spectrum 1-Spectrum 2)

technique can be applied to other separations where a variation in the relative amount of the structural components can be done.

This technique was first applied to determine the crystalline vibrational bands of trans-1,4-polychloroprene [54]. The spectrum of a cast film of predominantly (>90%) trans-1,4-polychloroprene polymerized at —20 °C was compared with the same sample heated to 80 °C (above the melting point) for 15 min. (Fig. 5). Elimination of the amorphous contribution of the composite semicrystalline spectrum was accomplished by subtracting spectrum b from spectrum a until the bands at 602 and 1227 cm^{-1} were reduced to the base line. The "purified" crystalline spectrum is given by spectrum c at the top of (Fig. 5) and exhibits the sharp band structure expected for a regular crystalline array. The interesting aspect of the crystalline spectrum was the observation that when the crystalline component spectra were obtained for samples polymerized at different temperatures, through the same procedure, the crystalline vibrational frequencies were different [55]. This should not be the case if the crystalline phase had the same structure. However, the spectra indicated that structural defects were being imbibed into the crystalline domains and there were more defects as the polymerization temperature increased. As a result of the defects occuring in the crystalline phase, the structure of the crystalline phase was different and consequently the spectra reflected this difference. However, to date, this is the only crystalline spectrum showing this effect, yet it was the first observed!

The separation of the crystalline and amorphous phases into their respective spectra has been carried out for a number of polymers including polyethylene terephthalate [56], polystyrene [57], poly(vinyl chloride) [58], polyethylene [59,60] nylon [61], polypropylene [62], and poly(vinylidene flouride) [63].

There are two fundamental assumptions in the use of absorbance subtraction. First, the absorbance and shape of a band does not change with the optical thickness. This assumption is tested with every subtraction and if the residual absorbance after a subtraction has a different shape than the original absorbance bands, then this assumption is violated and the procedure or samples should be reexamined for the appearance of effects producing nonlinear results. A number of these effects will be discussed below. Secondly, one also assumes that the absorbance of a mixture is the sum of the absorbances of the components, that is, that the components do not interact with each other differently at different relative concentrations, such interactions will lead to frequency shifts and band shape changes [75], but rarely are such effects observed in solids. Care must be exercised, however, since wedging and orientation can produce nonlinear effects in solids.

Because of the high sensitivity of absorbance subtraction to spectral differences it is to be expected that the technique is also sensitive to spectroscopic errors as well. It is probably not a misstatement to indicate that absorbance subtraction has generated more artifacts than facts due to lack of attention to the proper sample preparation procedures and spectroscopic techniques. For example if the sample is non-uniform or wedge-shaped, errors can occur [64,65]. A spurious difference spectra will be obtained between two spectra of the same sample at the same mean thickness if the samples are wedge-shaped. The severity of the wedging effect depends on the method of determining the scaling factor and the optical thickness of the sample. Hirschfeld [65] has proposed a method of correcting the spectra if the magnitude of the thickness change is known. However, since the thickness variations in polymer films are seldom uniform, the correction method can only be applied to solutions in wedge-shaped cells. However, if the optical thicknesses are kept below an absorbance of 0.8, the magnitude of the effect is minimized. When wedge-shaped samples are suspected, the scaling factor should be calculated using integrated peak areas to obtain the best results.

When the samples are highly uniform in thickness with smooth surfaces, another irritating effect can be observed and that is the presence of interference fringes. FT-IR offers some possibilities for eliminating fringes if the "signature" can be detected in the interferogram. This signature appears as a single spike and can sometimes be removed from the interferogram and consequently from the spectra [65]. Another approach is to use absorbance subtraction itself to remove the fringes. Since the reflection interferes to give a sine wave in the spectrum, one can generate with the computer a file containing a sine wave of the same amplitude and phase as the experimentally observed curve and subtract the computer generated fringes from the observed fringes. Another problem often encountered in absorbance subtraction is preferential orientation of one of the samples relative to the other. The inherent polarization of the beam is substantially less in FT-IR than in dispersion spectrometers where the polarization arises from the beam splitter while in dispersion spectrometers the gratings introduce substantial polarization. Unfortunately, there is no general method or algorithm for eliminating this orientational problem since each band has its own characteristic dichroic behavior [67]. The only solution is to make a determined effort to prepare the samples reproducibily. Unless this is accomplished one must resort to three dimensional tilted sample methods [68,69] for removal of the orientation and although spectral subtraction is helpful for carrying out this process [70], the removal of the orientation and the calculation of structure factors eliminates the need for absorbance subtraction.

Other experimental aspects of absorbance subtraction as related to the use of FT-IR spectrometers have been discussed including resolution [71] and instrument line shape effects [72], wave-length scale shift due to vignetting [73], noise and finite register effects [74] formation of "tertiary" interferograms due to multiple reflections [76] and others [77]. The discussion by Hirschfeld should be examined since he presents the most indepth discussion of the many possible instrumental errors and methods of eliminating them, where possible [77]. The above discussions have dealt with those absorbance subtractions where no shift in frequency occurs. There are circumstances where the spectral differences are associated with frequency shifts arising from changes in conformation, molecular interactions, etc. In these cases, absorbance subtraction can be used to accurately measure the magnitude of the frequency shifts [78,79]. If an absorption band has its center shifted by a fraction of its half-bandwidth and the shifted band is subtracted from the unshifted band, the absorbance difference ($A_1 - kA_2$) is a nonlinear function of the center frequency. The center frequency shift is determined from the measured absorbance difference by comparison with the difference of computed Lorentzian functions [78] or Gaussian functions [79]. Hence very small frequency shifts associated with change in the hydrogen bonding of sulfhydryl groups in hemoglobin can be detected, making the sulfhydryl groups a probe of structural alteractions at known locations within the hemoglobin molecule [78]. Similarly, the small frequency shifts associated with stressed polymers can be accurately measured [79].

The applications of absorbance subtraction to the study of polymers are numerous as later sections of this review will reveal. The uses of absorbance subtraction for identification [80,81], quality control [80,82] and forensic applications [83] have all been reported.

FT-IR has been used to determine residual catalyst support in commercial polyethylene at the level of 100 parts per million [84]. The method is based on the use of the 1118 or 470 cm^{-1} bands to determine the quantity of silica support dispersed in the polymer. The band at 2020 cm^{-1} was used as an internal standard for the amount of polyethylene. Neutron activation analysis was used to calibrate the weight percent of silica present in each polymer sample.

4.2 Ratio Method

One of the major problems in studying polymers quantitatively is the absence of model compounds for the purpose of calibration. A method of obtaining spectra of the components of a mixture spectra is based on obtaining the ratio of absorbances. This method was first used by Hirschfeld [85] for mixtures of components differing in relative concentration. This approach was later generalized but is limited to a rather small number of components since otherwise it is difficult to sort out the various ratios associated with each component [86]. Even more recently a careful look has been taken at the limitations of the absorbance ratio method [87].

The infrared spectrum of a two-component mixture can be represented by

$$M(v) = f_1(v) + f_2(v)$$

where $M(v)$ = spectrum of a mixture of components and $f_1(v)$ = spectrum of a pure component 1. The spectrum of a mixture of the same two components in different proportions is represented as

$$M_1(v) = a_1 f_1(v) + a_2 f_2(v)$$

Solving the equations for $f_1(v)$ and $f_2(v)$ one obtains

$$f_1(v) = \frac{1}{a_1 - a_2} M_1(v) - \frac{a_2}{a_1 - a_2} M(v)$$

$$f_2(v) = \frac{1}{a_2 - a_1} M_1(v) - \frac{a_1}{a_2 - a_1} M(v) .$$

The ratio spectrum

$$R(v) = \frac{M_1(v)}{M(v)} = \frac{a_1 f_1(v) + a_2 f_2(v)}{f_1(v) + f_2(v)}$$

defines the coefficients a_1 and a_2 by means of "flat areas". In a spectral region where $f_1(v) \gg f_2(v)$, $R(v) \approx a_1$. Conversely, if $f_2(v) \gg f_1(v)$ in any region, $R(v) \approx a_2$. The accuracy to which $f_1(v)$ and $f_2(v)$, the pure component spectra, are calculated depends solely on the accuracy to which a_1 and a_2 are determined. Computer simulation studies indicate that the determination of the coefficients is inaccurate if band overlap occurs or if small frequency shifts occur with the differing concentrations in the mixtures. The advisable method is to seek the maxima and minima in the spectra and calculate the spectra. Mixing of the spectra will be revealed by apparently negative absorbances in each of the component spectra. It is then necessary to iterate the values of the coefficients until the calculated spectra are "unmixed". It is to be observed that the calculated spectra are not the spectra normalized to unit concentrations or molarity, but contain a relative concentration term which depends on the concentrations of the initial mixtures used to calculate the spectra. It is necessary to carry out an "internal calibration" to obtain the specific or molar absorptivities [88]. This method has proved feasible in deriving spectra of the pure components from the spectra of two component mixtures. Care must be taken to properly account for band overlap and band shift problems. The method has the power to calculate spectra of "pure" component spectra for systems where the "pure" components cannot be modeled by suitable standard compounds. In particular, for polymer applications the spectra of a "pure" crystalline and a "pure" amorphous polymer can be obtained whereas in reality such pure model systems cannot be prepared. It is also possible to calculate the spectra of pure rotational isomers of polymer chains whereas in reality such pure systems have not been prepared [89]. This method has been successfully used to obtain the spectra of the pure gauche and trans isomers of poly(ethylene terephthalate) [90].

4.3 Factor Analysis

Factor analysis is based upon expressing a property as a linear sum of terms called factors. The technique has found wide application to a variety of multi-dimensional problems [91,92,93,94,95]. The technique has been applied to infrared and Raman spectra and most recently to FT-IR spectra [96].

The Beer-Lambert law can be written for a number of components over a wave length range as

$$A_i = \sum_j^n \varepsilon_j c_{ij}$$

where A_i is the absorbance spectrum of mixture i, ε_j is the absorptivity for the jth component, and c_{ij} is the concentration of component j in mixture i. Factor analysis is concerned with a matrix of data points. So in matrix notation we can write the absorbance spectra of a number of solutions as

$$A = EC\,,$$

where A is a normalized absorbance matrix which is rectangular in form having columns containing the absorbance at each wavenumber recorded and the rows corresponding to different mixtures being studied. The A matrix could thus be 400 by 10 corresponding to a measurement range of 400 wavenumbers at one wavenumber resolution for 10 different mixtures or solutions. E is the molar absorption coefficient matrix and conforms with the A matrix for the wavelength region but only has the number of rows corresponding to the number of absorbing components. C is the concentration matrix and has dimensions of the number of components by the number of mixtures or solutions being studied. Of course, we do not know E and C or we would not have a spectroscopic problem since E and C contain all of the information required to interpret A. In principle, factor analysis can be used to generate E and C which will allow a complete analysis of a series of mixtures containing the same components in differing amounts [97].

There are two basic assumputions in factor analysis. First, the individual spectra of the components are not linear combinations of the other components and second that the concentration of one or more species cannot be expressed as a constant ratio of another species. It is the different relative concentrations of the components in the mixtures that provides the additional information necessary to deconvolute the spectra.

What factor analysis allows initially is a determination of the number of components required to reproduce the adsorbance or data matrix A. Factor analysis allows us to find the rank of the matrix A and the rank of A can be interpreted as being equal to the number of absorbing components. To find the rank of A, the matrix A^TA is formed where A^T is the transpose of A. This matrix, termed the covariance matrix, has the same rank as A but has the advantage of being a square matrix with the dimensions corresponding to the number of mixtures being examined. In the absence of noise, the rank of A is given by the number of non-zero eigenvalues of M.

Since the actual data contains noise and computational roundoff errors, additional nonzero eigenvalues (noise eigenvalues) will be generated by the computation. The theory shows that the eigenvalues can be grouped into two sets a set which contains the factors or components together with an error contribution and a secondary set composed entirely of error.

It can be shown that

$$\sum_{i=1}^{r} \sum_{k=1}^{k=p} (A_{ik} - \hat{A}_{ik}) = \sum_{j=n+1}^{j=p} \lambda^{\circ} = \sum_{i=1}^{i=r} \sum_{j=n+1}^{j=p} \sigma_{ij}^2 .$$

The residual standard deviation (RSD) is

$$(RSD)^2 = \sum_{i=1}^{r} \sum_{j=n+1}^{p} \sigma_{ij}^2 / r(p - n)$$

where n is the number of components, p is the number of mixtures and r is the number of absorbances (number of absorbances reported for the wavenumber range scanned). Thus one can calculate the RSD and compare it with the experimental error. One of the recurring problems in factor analysis is deciding whether a calculated eigenvalue is associated with a true component or an error eigenvalue. Statistical tests have been developed to try to distinguish between the components and the error eigenvalues [93]. These tests should be carried out and compared with each other before a final decision is made. Shurvell favors recalculating the spectra and testing the goodness of fit [92] resulting from the addition of each successive eigenvalue ordered with respect to their magnitude. The goodness of fit parameter, Φ, is given by

$$\phi = \sum_{i} T_i^2$$

where

$$T_i = (T/T_0)_i \text{ obs} - (T/T_0)_i \text{ calcd.}$$

The goodness of fit parameter will approach a minimum which should correspond to the real error in the absorbance data and increase when a "noise" eigenvalue is added. Another test is the "indicator" function which also passes through a minimum when the correct number of eigenvectors are used. The "indicator" function is given by

$$IND = (RSD)/(p - n)^2 .$$

As an example of the results obtained on a real system, let us consider the case of determining the spectra of compatible polymer systems. If the polymers are incompatible when mixed, the resulting blend spectra should contain only two components since the polymers will phase separate and exhibit the spectra associated with the isolated polymers. For a compatible system, the intermolecular interactions leading to the compatible system may produce a spectrum associated with the "complex" between the two polymers so that three components might be expected. The spectra of a series of blends of polyphenylene oxide (PPO) and atactic polystyrene which are compatible have been studied [100]. For comparison, blends of polystyrene and chlorinated PPO which are incompatible were also studied. The IND function was calculated as shown in Table 1 for the two systems. For the compatible blend, the indicator function had a minimum corresponding to three components while for the incompatible system the minimum corresponded to two components in agreement with our expectations. By this process, not only has the number of components been deduced but also the experimental error involved in the measurements.

The determination of the spectra and the relative concentrations of each component in the mixture spectra are the results that make factor analysis worth the effort. This statement appears almost too good to be true but, with some restrictions to be noted later, it is possible. Let us begin with the covariance matrix

$$M = A^T A$$

which was diagonalized by finding a matrix Q such that

$$(Q^{-1} MQ)_{ij} = \lambda_j \delta_{jk}$$

Table 1a. PPO/PS Compatible System

Files	Component	Eigenvalue	Log (Eigenvalue)	Ind · 10^5
PS1	1	118.176533	2.07467945	40.0
PS2	2	5.196162	0.71568272	12.2
PS3	3	0.223502	−0.65071753	6.7
PS4	4	0.015321	−1.81469834	8.0
PS5	5	0.011154	−1.95254725	9.7
PS6	6	0.006483	−2.18821889	11.8
PS7	7	0.001856	−2.73128631	22.4
PS8	8	0.001369	−2.86329026	53.7
PS9	9	0.000299	−3.52361626	—

Table 1b. PPO/P4ClS Incompatible System

Files	Component	Eigenvalue	Log (Eigenvalue)	Ind · 10^5
CL1	1	187.03021	2.27191178	84.8
CL2	2	12.256074	1.08835140	18.8
CL3	3	0.137650	−0.86122316	21.3
CL4	4	0.094318	−1.02540278	22.4
CL5	5	0.038700	−1.41228050	24.2
CL6	6	0.009875	−2.00543276	38.5
CL7	7	0.003355	−2.47427252	123.0
CL8	8	0.001580	−2.80127694	—

where the set of λ_j eigenvalues of the set of equations

$$MQ_j = \lambda_j Q_j$$

and Q_j is the jth column of the matrix Q and δ_{jk} is the Kronecker delta function. The columns of Q are the set of eigenvectors M. The matrix of eigenvectors, Q, is a representation of the matrix C and the matrix V is a representation of the matrix E.

$$Q^T \equiv C\,, \qquad V = AQ \equiv E\,.$$

Q^T and V are representative of C and E, respectively in that, after the proper number of factors are chosen and the proper number of eigenvectors are incorporated in Q^T, Q^T has the same dimensions as C and V the same as E. Finally V multiplied on the right by Q^T reproduces the data matrix

$$VQ^T = A\,.$$

So one can consider V as the eigenvector representation of E and Q^T as the eigenvector representation of C. Unfortunately, these matrices in their present form have little physical significance. The matrix Q^T whose rows represent eigenvectors of the covariance matrix must necessarily be orthogonal to each other and therefore must be negative at some points. The concept of a negative concentration is novel but unacceptable.

The trace C matrix is, therefore, equal to a non-orthogonal rotation of Q^T

$$C = Q^T R$$

where R is a non-orthogonal rotation matrix. The matrix E must be obtained by use of the so-called right pseudo-inverse of C.

$$E = AC^T(CC^T)^{-1}$$

Thus the problem of deducing E and C is the determination of the proper non-orthogonal rotation matrix, R.

A number of different approaches have been taken to determine the rotation matrix [93,98,99,101]. The approach relies on the existence of unique frequencies for each component [98]. The procedure for finding the unique frequencies takes advantage of their uniqueness in their relative amounts of eigenvector composition. First, the V matrix is normalized so that the sum of squares across each row equals one. This places the eigenvector representation of the data on the surface of a hypersphere of dimensionality n. This is done so that only the directions or relative amounts of eigenvectors making up the frequency representations are considered. The unique frequencies will then bracket all the other representations in the eigenvector space. This first unique frequency point is the frequency representation containing the least relative amount of the first eigenvector, V(i1). The second unique frequency point is chosen by the one whose difference is greatest between it and the first unique frequency point and this value is chosen for the second eigenvector. The third is chosen by the greatest difference between it and the average of the previously chosen unique frequencies. If the unique frequency points are correctly chosen by the procedure described, the relative amounts of the eigenvectors required to represent the pure spectra will have been found. That is, the n rows of the normalized absorbances in the V matrix-which are chosen as the n frequencies with the most unique behavior, and provide the columns of the non-orthogonal rotation matrix R. The columns of the non-orthogonal rotation matrix R are taken to be the values across the rows of the V matrix, which is normalized so that the sum of the squares of the absorbances equals one. The rotation matrix allows the calculation of the relative contributions of each of the original mixtures to the spectra of the pure components. Figs. 6 and 7 show the result for a computer simulated system consisting of four mixtures of two different bands. The intersection of extrapolated straight lines with the zero-absorbance curve yields the coefficients required to calculate the pure spectra [101]. The method has been used to generate the pure spectra of the components of an acetylene-terminated monomer [101]. Factor analysis can also be used to reduce the noise in the

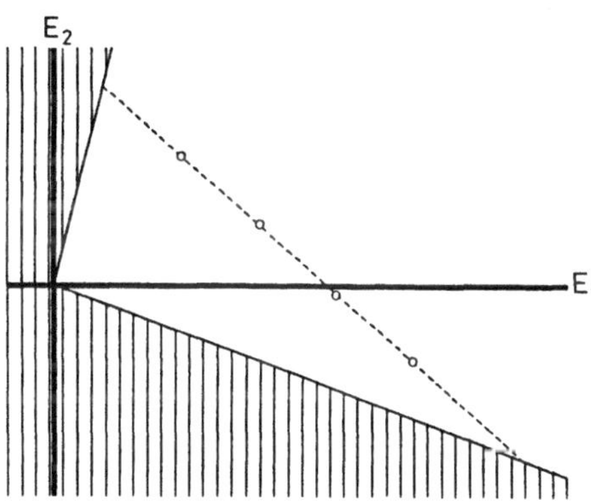

Fig. 6. A plot of eigenvectors E_1 and E_2 with the shaded areas showing regions of negative absorbances the data points represent the four mixtures: The extrapolated points represent the contributions of the two eigenvectors to the "pure" component spectra

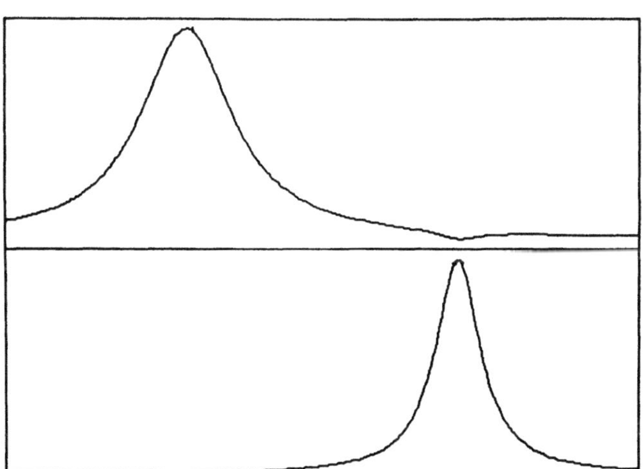

Fig. 7. Extracted pure component spectra from Lorentzian model mixtures

reconstructed spectra of the original components [102] and also as a method of searching spectral libraries for spectral identification [103]. Factor analysis has been applied to the analysis of the in-situ composition of epoxy resins in composites [104,105].

4.4 Least Squares Curve Fitting for Quantitative Analysis

Classically, quantitative infrared analysis was carried out using a single analytical frequency characteristic of the structure being determined in polymers. Least-squares regression analyses have been developed for utilization in the determination of multicomponent mixtures with overlapping spectral features [106]. A least-squares analysis uses all of the spectral data in the region of interest. The inclusion of all of the data in the spectral region substantially improves the precision and accuracy of the results [107]. The newer programs allow the simultaneous determination of the baselines as well as the spectra. It is also possible to set a threshold value below which the data will not be used thus using only those regions of the spectra with spectral information [108]. Weighting factors can be used to make maximum use of the spectral data available and minimize those regions with high noise response [108]. One of the advantages of the least-squares techniques is that no assumption is made about the spectral line shapes and this aspect of the technique makes it particularly useful for polymer analysis where the band shapes are asymmetric. It is possible to use the least-squares technique to do band shape analysis [109] and to establish non-subjective criteria for absorbance subtraction [110].

5 Experimental Techniques in FT-IR

The optics of the sampling chamber of commercial FT-IR instruments are the same as the traditional dispersive instruments. The sampling accessories which are generally available commercially can be used with FT-IR instruments. The main difference between the two types of instrumental optics is that the beam is round and larger at the focus for FT-IR. Thus sampling accessories may block some of the beam energy in FT-IR experiments. When energy is a limiting factor, the accessories can be modified to accommodate the larger beam.

However, the improved sensitivity of FT-IR allows one to obtain better sensitivity using the conventional sampling accessories and expand the range of sampling techniques. Emission, diffuse reflectance and photoacoustic spectroscopy represent new areas where FT-IR reduces the difficulty of the techniques considerably. Greatly improved results are also achievable from reflection spectroscopy. Special effects such as vibrational circular dichroism can be observed using FT-IR instrumentation.

Figure 8 shows diagrams of the various sampling techniques used in the infrared studies of polymers [111]. The principalle techniques for use will be discussed below.

5.1 FT-IR Transmission Measurement of Polymers

For successful application of FT-IR to polymer problems, sophistication in the methods of sample preparation are required in order to make the most use of the

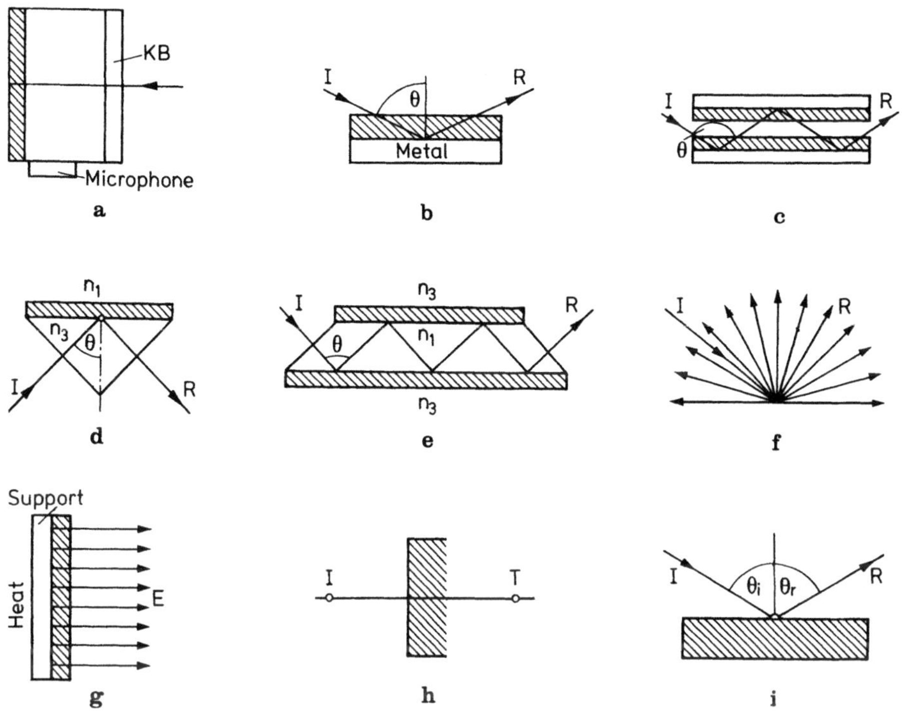

Fig. 8a) PAS cell, the incident light produces pressure fluctuations which are detected by a sensitive microphone; **b)** Single reflection RA set-up. Light penetrates the sample first and is reflected by the metal mirrors (θ should be 70 to 89.5°); **c)** Multiple reflection RA set-up. Light penetrates the sample first and is reflected by the metal mirrors (θ should be 70 to 89.5°); **d)** Single reflection IRS set-up. Light passes through the IRE first and is totally reflected at $\theta > \theta_c$; $n_1 \sin \theta_c = n_2 \sin 90°$; $\sin\theta_c = n_2/n_1$; **e)** Multiple reflection IRS set-up; **f)** Diffuse reflectance, the scattered light is collected by mirrors and directed to the detector; **g)** Emission technique, the sample is heated and the emitted radiation is analyzed; **h)** Transmission spectroscopy; **i)** Spectral reflection (mirror-like), angle of incidence equals angle of reflection

instrumental capabilities. The necessary methods have been described [112] and the principal difference from the traditional methods is the desirability of preparing larger sized (in diameter not thickness) samples so that the larger beam and its corresponding larger energy throughput can be used to full advantage. Solvent-casting or compression molding of polymer films yield high quality films but the thermal and processing history of the sample is lost.

For liquids and solutions, a 10% energy throughput advantage can be gained by machining the windows wider than the commercial liquid cells. This additional energy throughput is useful for strongly absorbing solvents like water [113]. For aqueous solutions, cell windows of PbSe or PbS can be used to prepare the cells of 6–15 micrometers in pathlength.

When insufficient sample is available, micro-sampling becomes necessary. Preparation of a KBr micro-disk should take into account sample beam geometry such that all the sample is in the beam, allowing high sensitivity. An alternate to the KBr disk is the diamond cell [114]. The commercially available cell (High Pressure Diamond Optics) has two diamond anvils made of type IIA diamond. The sample is placed between the two flat surfaces of the diamonds which are cemented in place in hollow pistons. A screw allows one to close the diamonds and develop a pressure to hold the sample or to study the effects of pressure [116]. The diamond cell can be used to study fibers, fragments of plastics, etc. The diamond cell can be used without a beam condensor [115].

Recently a detailed study has been made of the methods required to obtain spectra of micro and ultra-micro amounts of sample [117]. A focussing micro-illuminator is used for the detection of 1–10 micrograms of samples. Often, one can obtain the spectrum of impurities like "fish-eyes" in plastics by merely punching a hole the size of the imperfection in a masking material like cardboard and run the spectrum until the signal-to-noise level has reached a sufficient value to make an identification of the impurity. One can carry out a similar measurement in regions where the impurity appears to be absent and subtract the two spectra to enhance the spectra of the impurity relative to the more dominant polymer.

Textile fibers have always presented a problem for the infrared spectroscopist particularly if one does not want to modify the structure by the sampling technique. The higher dynamic range and the data processing capability have helped, but the problem still reduces to proper sample preparation. The most recent recommendation for FT-IR [118] has been to lay the untextured yarns on a salt plate, add a small amount of paraffin oil, and tease the fibers under a low power microscope with a tweezers and a fine brush until they are nearly parallel and a minimum of light leakage occurs between the fibers. For textured fibers, the continuous winding technique is perhaps best [119].

5.2 Diffuse Reflectance Spectroscopy

When light is directed onto a sample it may either be transmitted or reflected. Hence, one can obtain the spectra by either transmission or reflection. Since some of the light is absorbed and the remainder is reflected, study of the diffuse reflected light can be used to measure the amount absorbed. However, the low efficiency of this diffuse reflectance process makes it extremely difficult to measure [120] and it was speculated that infrared diffuse reflection measurements would be futile [120]. Initially, an integrating sphere was used to capture all of the reflected light [121] but more recently improved diffuse reflectance cells have been designed which allow the measurement of diffuse reflectance spectra using FT-IR instrumentation [122].

The requirement, by definition, for reflectance to be diffuse is that the intensity of reflected light is isotropic but for a powdered sample both scattering and absorption occur, and since the scattered radiation is angularly distributed, it may not be isotropic. However, with a sufficiently large number of particles, as is found in a powder, an isotropic scattering distribution can be achieved, so the emerging light will still be diffuse [123]. The Kubelka and Munk theory relates the function $f(R_\infty)$ [141, 142, 143] to the absorption coefficient (k) and the scattering coefficients (s).

$$f(R_\infty) = \frac{(1 - R_\infty)^2}{2R_\infty} = k/s$$

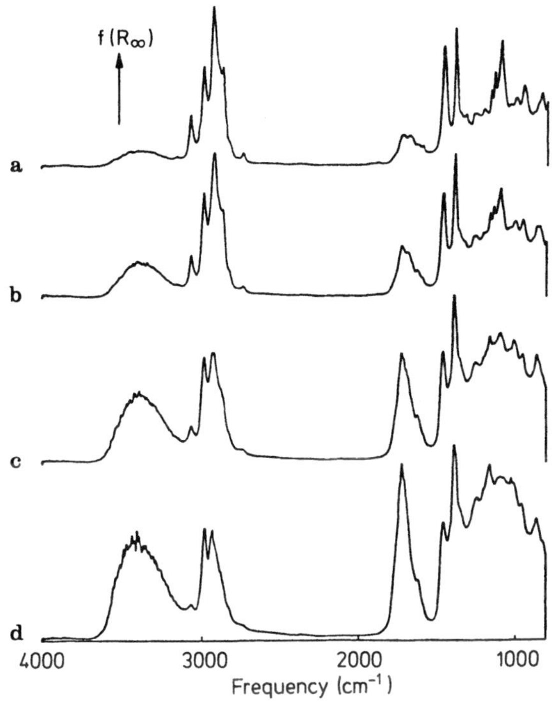

Fig. 9a—d. Diffuse reflectance spectra of a sample of powdered poly(dimethylfulvene) as a function of time; **a**) 15 min; **b**) 70 min; **c**) 255 min. and **d**) 1285 min. after sample preparation

where R_∞ is the absolute reflectance of an infinitely thick layer. In practice a standard is used and the following ratio is calculated

$$R' = R'\,(\text{sample})/R'\,(\text{standard})$$

where finely ground KBr has been recommended as a diffuse reflectance standard [122]. Figure 9 shows the diffuse reflectance spectra of a sample of powdered poly(dimethylfulvene) as it is oxidized in air as a function of time. The oxidation can easily be followed in this manner [122].

The principle problem with diffuse reflectance is that the specular component of the reflected radiation, that which does not penetrate the sample, is measured along with the diffuse reflected light which penetrates the sample. Generally, the change in specular reflection with frequency is small except in regions of strong absorption bands where the anomalous dispersion leads to Reststrahlen bands in the specular reflection spectrum. When the Reststrahlen bands are observed, the absorption bands can appear inverted at their center. This effect makes quantitative measurements on samples with strong absorptivity very difficult.

For powdered samples, diffuse reflectance offers considerable advantage particularly since no sample preparation is required which could change the morphology of the sample.

5.3 Internal Reflection Spectroscopy

In internal reflection spectroscopy (IRS) the sample is in optical contact with another material (e.g. a prism). The prism is optically denser than the sample. The incoming light forms a standing wave pattern at the interface within the dense prism medium, whereas in the rare medium the amplitude of the electric field falls off exponentially with the distance from the phase boundary. If the rare medium exhibits absorption, the penetrating wave becomes attenuated, so the reflectance can be written

$$R = 1 - kd_c$$

where d_c is the effective layer thickness. The resulting energy loss in the reflected wave is referred to as attenuated total reflection (ATR). When multiple reflections are used to increase the sensitivity, the technique is often called multiple internal reflection (MIR). Thus qualitatively, an IRS spectrum resembles a transmission spectrum. There are two adverse effects arising from the wavelength dependance of IRS. First, the long wavelength side of an absorption band tends to be distorted and second, bands of longer wavelengths appear relatively stronger. Fahrenfort was the first to demonstrate the usefulness of the phenomenon of attenuated total reflection (ATR) [124, 125] or total internal reflection, while Harrick developed the technique [126, 127] and designed ATR cells for commercial use.

With FT-IR spectrometers, one does not achieve the same improvement in IRS as in transmission compared to dispersion instruments because the ATR attachments have not been redesigned for the larger round beam. However, the signal averaging capability and speed have increased the utility of IRS for polymers [128] particularly for surface studies. In IRS, the infrared beam penetrates the surface of the polymer between a few tenths of a micron to a few microns depending on the type of reflection plate, the angle of incidence, and the wavelength of the infrared beam. The depth of beam penetration can be reduced by placing a thin barrier film between the trapezoidal reflection plate and the polymer under study [129]. For the study of the surfaces of polyurethanes a barrier film of 2,5-dichlorobenzotrifluoride was used since its infrared absorption does not interfere with the polyurethanes. ATR dichroism studies for the purpose of characterizing polymer surfaces have been made [130]. The orientational character of the surface can be studied using a rotatable sample holder with a symmetrical, double-edged internal reflection crystal.

Hirschfeld [131, 132] has generated the algorithms which are necessary to use IRS to determine the optical constants of a sample from a pair of independent reflectivity measurements at each frequency. The optimum method is to determine the total reflectance at two polarizations at the same incidence angle.

5.4 External Reflection Spectroscopy

When the surfaces are highly reflecting as in the case of metals, external reflection spectroscopy (ERS) can be used with good success [133]. For optimum intensity of the reflection bands of thin films, angles of incidence near 88 are desirable. However, in order not to interfere with the incoming beam, angles of incidence near 80° are used.

In general for a highly reflective sample, 50 to 60 percent of the energy is lost through the reflection optics. If the optical constants of the film are known, the thickness of the film can be calculated based on optical theory [134,144]. Studies have been made of thin films of poly(methyl methacrylate) depolymerization on gold, nickel and zinc surfaces using ERS [135] and FT-IR.

Using ERS and an environmental cell which allows the simultaneous ultraviolet radiation under varying thermal and environmental stresses, the photodegradation of polycarbonate on gold mirrors was studied [145]. The solar induced degradation of polymers on metal surfaces can be studied. Factors such as metal surfaces, film thickness, atmospheric environment and U.V. source and intensity can be studied in this manner. For polycarbonate, exposures of the film to the solar simulator were made for various time intervals and the results compared. Increases in absorbance in the region 3500 to 3250 cm^{-1} and at 1690, 1620, 1590, 1490, 1340 and 1260 cm^{-1} suggest the formation of hydroxyl, a portion of which is hydrogen-bonded, a salicylate-like carbonyl, and a conversion of carbonates —C—O bonds to phenolic —C—O bonds. The results suggest the photo-Fries rearrangement is occuring for the polycarbonate [145].

5.5 Reflection-Absorption Infrared Spectroscopy

It is possible to increase the sensitivity of ERS by increasing the geometrical path length by making multiple reflections of the incident radiation at nearnormal incidence from opposing samples. This type of spectroscopy has been called reflection-absorption spectroscopy [136]. When light is incident on a highly reflecting metallic surface at near-normal incidence, the incident and reflected beams combine and form a standing wave having a node very near the surface of the metal. Accordingly the electric field of the radiation has zero amplitude at the surface of the metal and cannot interact with coatings on the surface [137]. However, this phenomenon can be avoided by using radiation polarized parallel to the plane of incidence and making an angle of incidence that is only a few degrees less than 80 [138]. In this manner, very thin films on the surface of a highly reflecting surface like a metal can be studied in the infrared region. Boerio and his collegues have used this technique in the dispersive instrumentation [139,140] to study silanes and epoxy resins on metal surfaces at a thickness of 15 Å. The reflected infrared intensities were less than 5%. The spectral results indicate that the molecules were absorbed on the iron and copper mirrors with a vertical conformation with, probably, only a single oxirane oxygen in contact with the surface [140].

5.6 Emission Spectroscopy

Kirchoff's law [146] states that a given temperature the absorbance, a, of a sample is equal to its emissivity, e. Therefore if one can measure the emission spectrum of a sample, it can be easily related to the absorbance spectrum [147]. An emission spectrum is usually measured from a hot sample to a colder detector, although the reverse can also lead to an emission spectrum. The magnitude of the emission signal is proportional to the fourth power of the temperature difference between the sample and the detector (Stefan's law). The major advantage of emission spectroscopy is that it does not require an infrared optical material to be in contact with the sample and in some cases, the spectra can be obtained "in situ" if the temperature of the sample is sufficiently high or low.

The emissivity is calculated from the measured emission by ratioing the measurement from a blackbody source at the same temperature as the sample [148]. Since there is a background emission from instrumental surfaces, four measurements are often made [149, 150], to remove the background emission. The measured intensity, S(v, T) at any temperature has several components

$$S(v, T) = R(v)\,[e(v, T)\,H(v, T) + B(v) + I(v)\,p(v)]$$

where v = frequency, T = temperature, R(v) = instrument response function; E(v, T) = emittance of sample; H(v, T) = Planck function; B(v) = background radiation; I(v) = background radiation reflected from sample; p(v) = reflectance of sample. For a blackbody reference, e = 1 and p = 0. Therefore, the intensities measured on the reference material are

$$S_1(v, T_1) = R(v)\,[H(v, T_1) + B(v)]$$

$$S_3(v, T_2) = R(v)\,[H(v, T_2) + B(v)]$$

similarly for the sample

$$S_2 = R(v)\,[H(v, T_1)\,e(v, T_1) + B(v) + I(v)\,p(v)]$$

$$S_4(v, T_2) = R(v)\,[H(v, T_2)\,e(v, T_2) + B(v) + I(v)\,p(v)]$$

and the emittance can be calculated from

$$e = (S_4 - S_2)/(S_3 - S_1)$$

Practically, the interferograms are substracted in order to avoid some phase correction problems. If one is working with a room temperature triglycine sulfate (TGS) detector, the four-measurement experiments are not necessary to correct the background [150], whereas with a liquid nitrogen cooled HgCdTe detector, the background correction is necessary.

When the temperature differential is high enough as is the case for molten inorganic salts at 600 °C, a dispersion infrared instrument can carry out the emission measurements [151], but dispersion instruments cannot be used for most organic and polymeric materials as the emission signals are too weak. However, because of the greatly increased sensitivity and lower noise levels of FT-IR, emission spectroscopy can be carried out with very small temperature differentials between the sample and the detector [152]. The first emission measurements made on FT-IR instrumentation were by Low and Coleman [153, 154] on a wide variety of materials including polystyrene and nylon. Bates and Boyd [155] applied emission FT-IR to record the spectra of molten salts. Griffiths [156] reported some early fundamental studies of the potential of emission spectroscopy by FT-IR. In principle, the emission spectra can be measured with the sample at room temperature using a low temperature detector, and this has been accomplished [157,158].

A number of general problems exist in emission spectroscopy and a number of articles have discussed potential solutions [159,160,161]. One of the problems has been termed self absorption, that is, the transmission and emission spectra of thin layers of materials correspond very well, but thick samples may shown continuous emission spectra of little value or spectra showing minima where maxima are expected. A pure emission spectrum can originate only from a material at uniform temperatures throughout. A temperature distribution through the material will lead to partial reabsorption of emission bands and a variety of band shapes can arise. The self absorption problem can be avoided by properly designing the heating cell so that heating is from the front so the outer layers are at the highest temperature. In fact, by varying the intensity of the heating radiation, different thicknesses of sample are probed [161].

A different situation arises if there are reflection or scattering effects which can contribute to the loss of transmission and to the form of the emission spectra. The selective reflections reduce the emission and occur in the vicinity of very strong absorptions [163]. These effects tend to reduce the relative intensity of the stronger emissions with respect to the weaker emissions.

One of the problems which must be solved for quantitative measurements by emission is the need for a blackbody source at the temperature of measurement. And a variety of blackbody references have been used including a V-shaped cavity of graphite [164], a metal plate covered with a flat black paint [156,160] and a cone of black paper [153]. However, none of these methods of producing a blackbody reference spectrum are adequate. In most cases the efficiency of the reference has not been established. The most recent recommendation [150] is an aluminium cup painted with an Epley-Parsons solar black lacquer which has an emittance of greater than 98% over the infrared spectral range.

The sensitivity of emission spectroscopy is related to the amount of sample and the temperature differential between the sample and the detector. However, one cannot increase the temperature of the sample to the point where sublimation or decomposition begins. Additionally thermal gradients in the sample must be minimized and the temperature must remain stable for the duration of the measurement, both of these factors are harder to control at higher temperatures. Another consideration for signal-to-noise ratios is the frequency dependence of the Planck blackbody radiation law. The energy peaks fairly sharply and decreases significantly on the high frequency side. Thus for any particular temperature, the relative noise will be greatest at the extreme ends of the spectrum. In practice, polymer samples are run between 70 and 130 °C. These temperatures limit the frequency range to 2000 to 450 cm^{-1}, with the 2000 to 1800 cm^{-1} region still fairly noisy. The bulk of the fingerprint region, however, retains good signal-to-noise ratios for emission spectra signal averaged over a thousand scans.

An obvious area of interest for emission spectrocopy is the area of coatings on metal surfaces [165]. Most metals give a very weak continuum emission of perhaps 5% of a backbody at the corresponding temperature. Some success has been reported in this area [149,152,160]. Emission spectra have been taken from the inside of Aluminium been cans which have been cured at different temperatures. Obvious spectral differences can be observed [165].

One of the most interesting uses of emission spectroscopy is the study of the action of lubricants [162,166,167,168,169,170]. A loaded steel ball is rotated in a fluid bath and made to slide over a diamond window. A contact region is formed which can be measured by the radiant power emitted. A substraction of the ball surface radiation from the total radiation emitted from the contact region was electronically carried out. A study of the emission spectra of a polyphenyl ether and a naphthenic fluid under dynamic conditions was made. The widths of some of the bands of the ether increased dramatically when a certain load (pressure) was exceeded. These increases were correlated with the changes of chemical composition through decomposition of the fluid [170,171].

The use of FT-IR emission will develop rapidly in the future as it has the potential to eliminate sample preparation entirely and perhaps allow one to take his polymer sample to the spectrometer and measure the spectrum directly.

5.7 Photoacoustic Spectroscopy

Photoacoustic spectroscopy (PAS) is a rapidly growing method of obtaining infrared spectra of samples that are hard to prepare as transparent films, or have high internal light scattering or are coated onto opaque or strongly light scattering substrates. The PAS technique is thus complementary to ATR and diffuse reflectance and has the advantage of no sample preparation. Typically, only 0.25 cc of sample are required and the specimen can usually be examined "as received". The signal-to-noise ratio for PAS is low, so a longer scanning time (1000–6000 scans) is required compared to transmission FT-IR.

The photoacoustic effect is simply the generation of an acoustic signal by a sample exposed to modulated light. The solid sample is placed in an enclosed chamber with a coupling gas such as air, helium or argon and is exposed to modulated light. The sample is heated to the extent that it absorbs the incident light and the energy gained is lost as heat through non-radiative processes. Because the light is modulated, the temperature rise is periodic at the modulation frequency, and it is this periodic temperature rise (typically <0.001 °C) at the surface of the sample that, in turn, causes a modulation of the gas pressure in the enclosed chamber. This pressure modulation is an acoustic signal and is detected by a sensitive microphone coupled to the chamber. PAS rises because a gas medium surrounding the surface of interest serves to act as a piston conducting the PAS signal from the solid surface to the microphone. If the solid sample in the PAS cell absorbs a particular infrared frequency, it will respond by generating an acoustic signal at the particular audio frequency corresponding to the incident infrared frequency. The result of absorption by the sample of several different frequencies is a PAS signal containing several different audio frequencies [172,173,174,175]. It should be immediately recognized that light reflected or scattered without any frequency translation cannot heat the sample so samples of this type which are difficult to examine by other methods are particularly suited to PAS.

The application of FT-IR instrumentation arises from the need for a high signal-to-noise and the multiplex advantage is helpful in this regard. The light output from the Michelson interferometer of the spectrometer is modulated at an audio frequency that is a function of the infrared frequency because the moving mirror arm of the interferometer contains a mirror moving with a linear velocity of approximately 0.3 cm/sec. The modulation frequency (f) is given by $f = vw$ where v is the velocity of the mirror and w is the frequency of the light. So the light is modulated at audio frequencies of <2 Khz in a Digilab instrument and changes an order of magnitude in going from 4000 to 400 cm^{-1}. In effect, a photoacoustic interferogram is produced. This signal is detected by the microphone. The output signal, after appropriate amplification, is input to the detector electronics and can be processed in the usual ways. The magnitude of the photoacoustic signal depends on a variety of factors, including the thermal and absorptive properties of the sample. It is primarily a function of the absorptivity of the sample at the frequency of radiation and the thermal diffusivity of the sample (the ratio of thermal conductivity to specific heat and density).

The first application of FT-IR-PAS was to gas samples [176)], although Low and Parodi [177,178,179,180,181)] used the technique in the infrared dispersion mode. Recently a number of papers have appeared using the PAS as an accessory on the FT-IR instrumentation [182,183,184,185,186)] and polymer applications are appearing [182,183,187,188,189)]. An interesting example is the comparison of the transmission FT-IR spectrum with FT-IR-PAS of an RTV silicone rubber loaded with SiO_2 powder and TiO_2. Conventional reflectance and transmission measurements could not provide the spectrum while the PAS method yields a clearly resolved spectrum showing all of the features of the infrared spectrum of the unloaded RTV. The PAS spectrum did not show features due to the SiO_2 and TiO_2; since these particles are encapsulated in a relatively thick RTV layer their PAS response is attenuated [189)].

In order to obtain quantitative information from a photo-acoustic spectrum, a number of factors must be taken into consideration [190)]. The magnitude of the PA signal depends on the intensity of the incident signal, the thermal and absorptive properties of the sample and the characteristics of the spectrometer itself. The signal depends monotonically on the absorption coefficient and obviously on the geometry, elastic and thermal properties of the sample, and the adjacent phases. The thickness or the depth of the sample penetration is on the order of 10 micrometers for typical polymers at 1000 cm^{-1} [191)]. This thickness represents an intermediate case when compared to ATR and transmission. The effective penetration depth can be varied by changing the light modulation frequency through a change in the interferometer mirror velocity [183)]. Band intensities decrease as the modulation frequency increases because the thermal diffusion depth is inversely related to the modulation frequency [192)]. Any sample whose optical absorption length is similar to or larger than the thermal diffusion length can be studied by PAS. The optical path length is $b_{op} - 1/a$ where a is the optical absorption coefficient. The thermal diffusion length is defined as $b_t = (2k/pCw)^{1/2}$ where k is the thermal conductivity, p = the density, C = the specific heat, and w is the modulation frequency. When $b_{op} \gg b_a$, one has an optically transparent and thermally thin solid, in which case the PAS signal is proportional to a and has an inverse dependence on the modulation frequency. Optically dense polymer samples can be studied by PAS, which probabily could not be studied by any other technique, unless the sample is made optically thin by fabrication, dilution or grinding into fine powders. Recently Krishnan [196)] compared the spectra of some polymers using PAS, transmission and diffuse reflectance and found that PAS

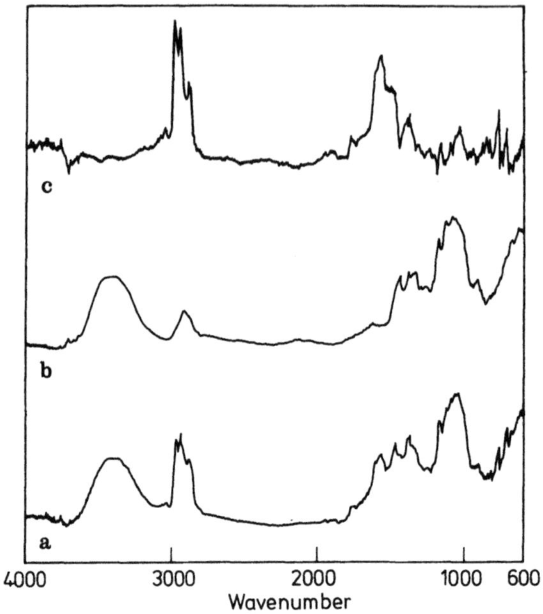

Fig. 10a—c. PA spectrum of a, paper with coating; **b)** paper without coating; and **c)** the difference **a** — **b**. Ref.: Fig. 9 from K. Krishnan, Appl. Spect., Vol. *35* (6) (1981)

and diffuse reflectance give very similar spectra. A particular application is the study of coatings on strongly absorbing substrates like paper. Figure 10 shows the PAS spectrum of an uncoated, coated and difference spectrum of paper [196)]. The PAS spectra are of sufficient quality for an identification. Depth profiling of surfaces has also been suggested using PAS [198)].

Another application is the reporting of the PAS spectrum of heavily n-dopped acetylene which is the prototype "organic" metal and for which the transmission spectrum or reflectance spectrum is very difficult to obtain [199)]. The fibrous nature of the material and its reactivity with air presented no problems for the gastight PAS sample cell.

For quantitative measurements, two factors become important, saturation and scattering [197)]. Saturation occurs when the optical absorption length approaches the thermal diffusion length. This problem is considered to be unimportant in the infrared region but a definitive study has not been made. The light scattering problem has recently been considered [193, 194)]. To quote Lloyd et al. "... a damning present criticism of PAS is that it is not yet a reliable quantitative analytical tool." [195)]. It is clear that PAS has a bright future in the application to polymer systems particularly if the quantitative aspects can be worked out.

6 Spectroscopic Techniques Using FT-IR

The data manipulating capability of a computerized infrared spectrometer allows the spectroscopist to delve more deeply into the structural origin of the infrared absorptions by using data processing techniques to purify, manipulate, and correlate the spectra. If one can systematically vary the relative amounts of various structural contributions, absorbance subtraction can be used to isolate the spectral contributions of the structural components.

6.1 Isolation of Structural Defects by Varying Polymerization Temperature

The problem encountered in attempting to detect the presence of structural irregularities is the considerable spectral band interference by the strongly absorbing predominant structures. Spectral subtraction allows the removal of these interferences. For example in polychloroprene, the predominant structure is the trans-1,4-polychloroprenes but there exist contributions due to cis-1,4,1,2 and 3,4-structural irregularites depending on the polymerization temperature. Coleman, et al. [200] isolated the spectral contributions of these minor structures by spectrally subtracting out (above the melting point of the polymers) the bands attributable to trans-1,4- units at 1660, 1305, and 825 cm^{-1}. (Fig. 11) shows the spectra at 70 °C for a polychloroprene polymerized at —20 °C (a) and at —40 °C (b) and the difference spectrum (b—a). The major bands in the difference spectrum are due to the cis-1,4 unit, (i.e. 1652, 1285, 1105, 850, 690 and 654 cm^{-1}). The resulting difference spectra allowed the detection of cis-1,4-polychloroprene units at the 4% level and at the 1% level for the 1,2- and 3,4-

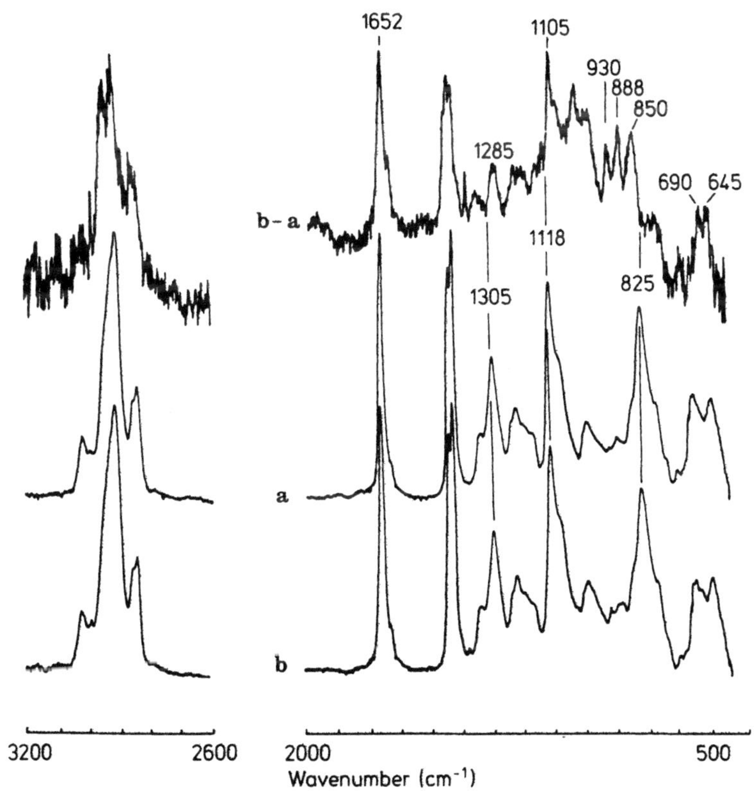

Fig. 11 a and b. Fourier transform i.r. spectra at 70 °C in the range 500–3200 cm^{-1}: **a)** polychloroprene polymerized at —20 °C; **b)** polychloroprene polymerized at —40 °C; **b—a)** difference spectrum.
Ref. Fig. 1 from M. M. Coleman, R. J. Petcavich and P. C. Painter, Polymer, 1978, Vol. 19, November

structural units. From the observation of these spectral bands arising from the structural impurities, band assignments could be made and subsequently used for the quantitative structural analysis of polychloroprene.

6.2 Isolation of Conformational Structures by Variation in Annealing Conditions

For many polymers with relatively simple chemical chain structures, the properties are dependant to a large extent on the conformational or rotational isomeric distribution in the polymer arising from its process or thermal history. Poly(vinyl chloride) PVC is such a polymer and it is possible to use absorbance subtraction to isolate the conformational sequences of PVC by varying the annealing process [201]. A rapidly quenched film was prepared by quenching into ice water from the melt at 200 °C. Subsequently, the sample was annealed at 80 °C cooled to room temperature and scanned. A similar process was carried out for annealing temperatures of 100, 120, 140 °C on the same film. In the difference spectra, positive absorbances reflected increases of the particular conformational sequence while negative absorbances reflected decreases. The observed changes were correlated with the theoretical predictions [202] and assignments made to the various conformational sequences such at TTTT, TTTG, and TTGG.

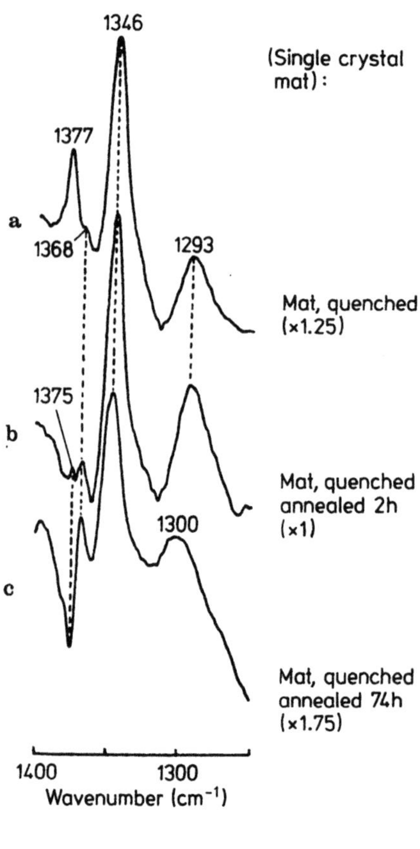

Fig. 12a—c. Single crystals. **a)** quenched, 1:0.65; **b)** quenched, annealed 2 hr, 1:0.95; **c)** quenched, annealed 74 hr, 1:0.98.
Ref. Fig. 4 from P. C. Painter, J. Havens, W. W. Hart, and J. L. Koenig, J. of Polym. Science, Polym. Phys. Ed., Vol. 15, 1223–1235 (1977)

Studies of annealing of polyethylene have revealed new insights into the structural assignments of some of the amorphous absorptions [209]. For example, a band at 1346 cm^{-1} appears when single crystal mats are compared with mats which have been quenched and subsequently annealed as shown in Fig. 12. The subtraction criterion is the reduction of the amorphous 1368 cm^{-1} band to the base line. The difference spectra obtained by substracting the spectra of quenched, quenched then annealed for 2 hr, and quenched then annealed for 74 hrs mats from the spectrum of the original mat are shown in (Fig. 12). In all three difference spectra the 1346 cm^{-1} band remains strongly positive. This observation indicates an absolute decrease in the 1346 cm^{-1} band upon quenching. Additionally the 1346 cm^{-1} band is not found in the spectra for extended-chain crystals indicating that the 1346 cm^{-1} band is associated with a conformation unique to solution-grown single crystals which is in addition attributed to the fold surface [209].

For polypropylene, by using the spectrum of an annealed sample and subtracting it from a quenched sample it is possible to obtain a difference spectrum characteristic of the amorphous regions of polypropylene [210]. In Fig. 13, the difference spectrum characteristics of the amorphous phase of the quenched sample (a) is compared with the difference spectrum characteristic of the ordered phase of an annealed

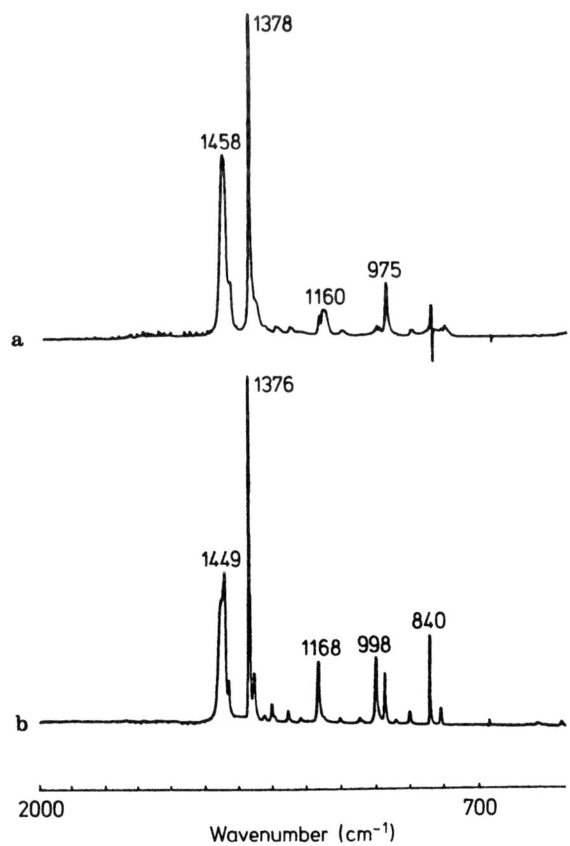

Fig. 13a. Difference spectrum characteristic of the amorphous phase of the quenched sample; **b)** difference spectrum characteristic of the ordered phase of the annealed sample.
Ref. Fig. 2 from P. C. Painter, Polymer, 1977, Vol. 18, November

polypropylene sample. The band at 1376 cm^{-1} was used as the criterion for absorbance subtraction. The difference spectrum of the amorphous phase is very similar to that of the melt. The features of the spectrum indicate that there are ordered helical chain segments in the amorphous phase of the quenched polypropylene samples.

The spectrum of crystalline isotactic polystyrene has been isolated by subtracting from each other the spectrum of a quenched (amorphous) and annealed semicrystalline film. The criterion for subtraction was the reduction of the amorphous band near 550 cm^{-1} to the baseline [211]. It is not precisely correct to term the resultant spectrum a crystalline spectrum since no effects due to interchain interactions have been isolated. It is more properly described as typical of long segments of helical conformations. The vast majority of such chains are found in the crystalline phase, (Fig. 14). The scale-expanded difference spectrum reveals new features. For the first time the doublets at 1365–1363 and 1303–1298 cm^{-1} are observed. The controversy over the origin of the doublet at 1083–1052 cm^{-1} is cleared up by an examination of a series of difference spectra resulting from annealing. The results suggest that the splitting is associated with the sequence length of the preferred conformation in the amorphous region rather than a rotational disorder in the orientations of the benzene ring in the crystalline regions.

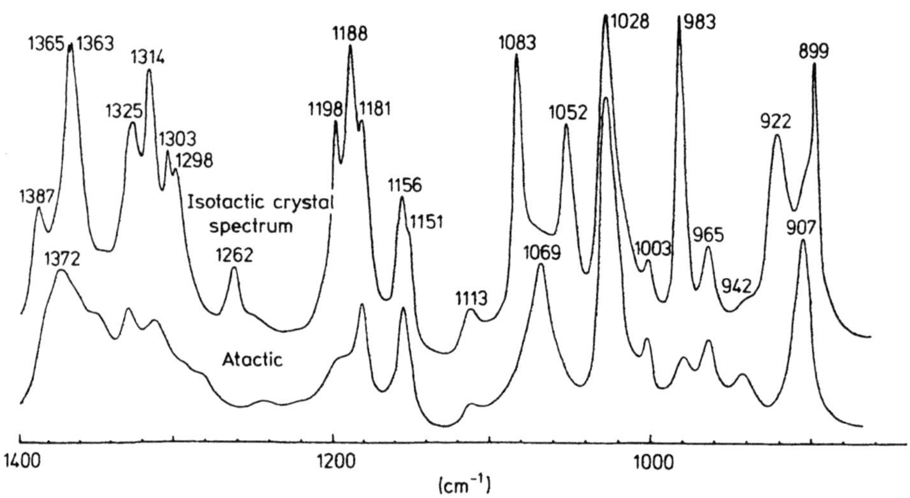

Fig. 14. Comparison of the crystal difference spectrum of IPS with the infrared spectrum of an atactic sample.
Ref. Fig. 3 from Paul C. Painter and Jack L. Koenig, J. of Polym. Sci., Polym. Phys. Ed., Vol. 15, 1885–1903 (1977)

Poly(vinylidene fluoride) $PVDF_2$, has been studied by absorbance subtraction in order to isolate the spectral features of the different phases; in particular, the difference spectra were used to interpret the structure of phase III [212]. The spectrum of the unoriented phase-III sample before annealing is shown in Fig. 15. The spectrum after annealing at 160 °C for 20 hr is also shown with the difference

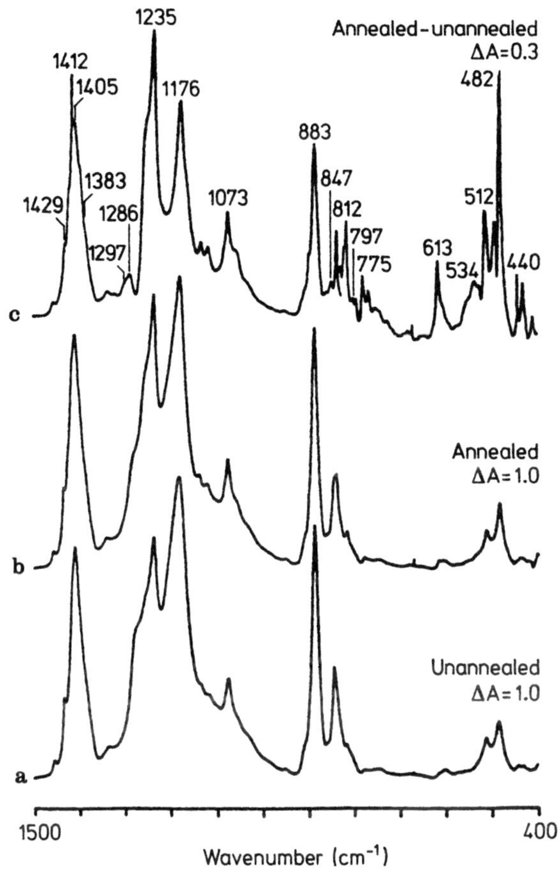

Fig. 15a—c. Phase-III PVF_2. **a**) asacast; **b**) annealed 20 h at 160 °C; **c**) (**b**—**a**) difference spectrum; ΔA = absorbance. Discontinuity at 825 cm^{-1} arises from the composite nature of (**c**).
Ref. Fig. 5 from M. A. Bachmann and W. L. Gordon, J. Appl. Phys., *50*(10), 1979

spectrum. From the change in absorbance of the bands at 1073 cm^{-1}, the crystallinity increase was 23 %. When the spectrum of the amorphous phase is removed, the crystalline spectra revealed approximately 45 bands as opposed to the 20 observed before subtraction in the annealed sample. The large number of bands of phase III rules out an all trans structure for the chains in phase III of PVF. Comparison of the spectrum of phase III with the spectra of phase II and phase I, reveals that there is a correspondence of the bands of the latter two phases in the spectrum of phase III. This correspondence of bands requires that the conformations found in phases I and II must be found in phase III. Utilizing the results from x-ray diffraction and potential energy calculations, the conformation TTTGTTTG′ was suggested for the chain structure of phase III [213].

Sometimes, small structural differences in morphology of polymer samples can be isolated by using a double subtraction technique. For example, with poly(ethylene terephthalate) PET, differences in the amorphous phase of the melt-quenched polymer and solution-cast polymer can be isolated by first subtracting out the contribution due to the trans isomer and then subtracting the two difference spectra from each other [214]. (Fig. 16) shows the resulting difference spectrum obtained after the second subtraction. Obviously the two amorphous structures are different from each other.

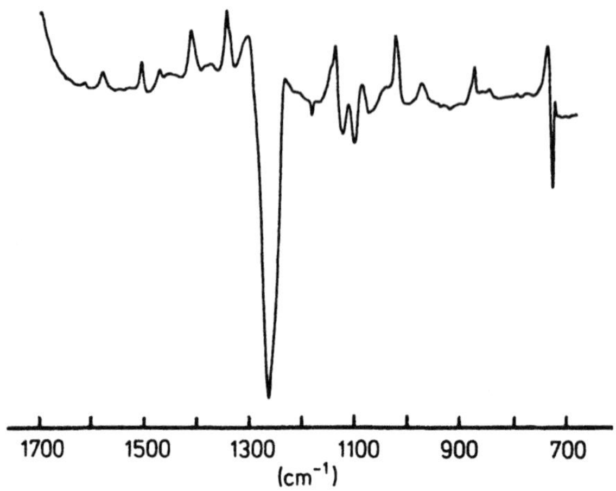

Fig. 16. Difference spectrum: melt-quenched amorphous component spectrum minus solution-cast amorphous component spectrum.
Ref. Fig. 6 from L. D'Esposito and J. L. Koenig, J. of Polym. Sci., Polym. Phys. Ed., Vol. *14*, 1731–1741 (1976)

6.3 Isolation of Conformational Structures by Varying Measurement Temperature

Infrared spectroscopy has often been used to measure energy differences between conformational isomers. With FT-IR one can systematically study the differences introduced by temperature by doing absorbance subtraction. Studies were made by examining the difference spectra of PVC recorded at elevated temperatures in the range of 80 to 180 °C [201]. From the intensities, Van't Hoff plots were made and energy barriers determined. These results further confirmed the band assignments to the various conformational sequences. Studies have also been carried out on PVC which has been plasticized [203]. In these studies the contributions of the plasticizer were substracted out to reveal the changes in the PVC conformations.

Conformational energy differences have been determined for isotactic, atactic and syndiotactic poly(methyl methacrylate) (PMMA) by FT-IR [205]. Difference spectra were obtained by subtracting the spectra at the elevated temperature minus the starting temperature. The peak absorbances at the characteristic frequencies as a function of temperature were analyzed using the Van't Hoff equation. The lowest energy conformation of PMMA is trans-trans and the high-energy conformation is trans-gauche. The conformational energies for the syndiotactic PMMA was 2000 cal/mol while a value of 700 cal/mol was found for isotactic PMMA and 1400 cal/mol for atactic PMMA. The FT-IR results are in general agreement with the calculated results using the rotational isomeric state theory [207]. This FT-IR data has also been used to test the Gibbs and DiMarzio [208] prediction that conformational energies are a primary factor in determining the glass temperature of a polymer. The Gibbs-DiMarzio theory predicts a constant value for the ratio of the conformational energies divided by glass transition temperature. This ratio is 2.6, 1.9, and 1.1 for the syndiotactic, atactic and isotactic PMMA, respectively. Clearly, this ratio is not constant and the differences probably arise from contributions due to the side-chains' conformations as well as the backbone conformations.

Differences in spectra obtained at different measurement temperatures were also used to aid in the determination of the structure of calcium and sodium ionomers [215].

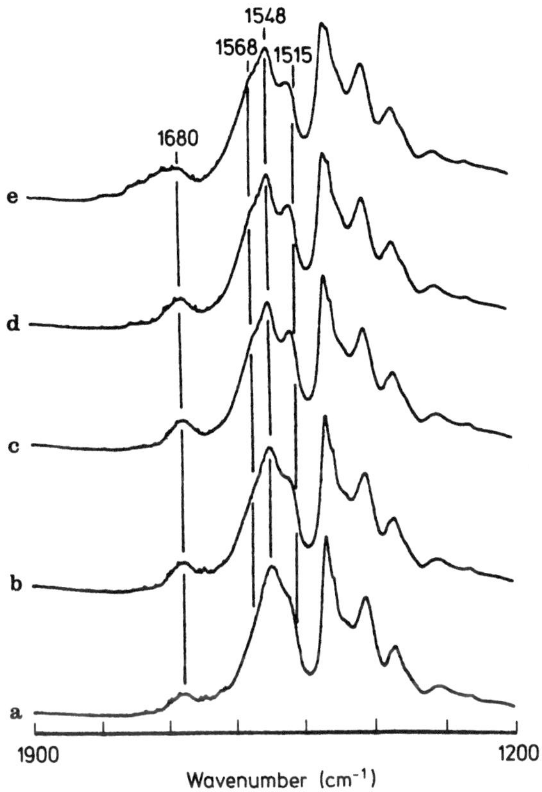

Fig. 17a—e. FTIR spectra in the range of 1900–1200 cm^{-1}: completely ionized calcium ionomer at (**a**) room temperature after quenching in liquid nitrogen from 190 °C; **b**) 40 °C after 30 min; **c**) 70 °C after 30 min; **d**) 130 °C after 15 min (after 15 min. at 90 °C); **e**) 150 °C after 5 min. Ref. Fig. 2 from P.C. Painter, B. A. Brozoski, and M. M. Coleman, J. of Polym. Sci.: Polym. Phys. Ed., Vol. 20, 1069–1080 (1982)

Figure 17 shows the infrared spectra (from 1900 to 1200 cm^{-1}) of the completely ionized calcium ionomer of an ethylene-methacrylic acid copolymer recorded at room temperature (a) and the same film recordered at 40, 70, 130, and 150 °C (b–e, respectively). This recording of the spectra at the elevated temperatures accentuates the sharp doublet at 1515/1548 cm which is characteristic of an interaction or vibrational splitting of the pairs of COO^- groups.

6.4 Isolation of Conformational Structures by Varying Applied Pressure

Another method of systematically changing the conformational distribution is through the application of pressure. FT-IR studies of the changes in the conformational states of PVC due to the application of pressure reveal that the high-energy conformations are frozen-in with pressure [204, 205]. Difference spectra were obtained from samples before and after application of pressure. At all pressures above 1 KBr the decreases in absorbances in the difference spectra were associated with decrease in trans conformation. Glassy samples prepared at high pressure show an excess of high-energy conformations (TG) which confirms that there is freezing of backbone conformations during glass formation. When the temperature and the pressure are varied, two regions of behavior are observed: (1) at low pressure, the rate of change of the pressure is large owing to the change in conformation with pressure; and (2) at high pressure, the rate of change of absorbance with pressure is small due to the change in the intermolecular interactions with increasing density.

6.5 Use of Isotopic Substitution in FT-IR

Isotopic substitution is a powerful molecular probe for making band assignments for the vibrational spectra of polymer molecules. For polymers where the absorbance bands have large band widths, there is often extensive overlap and there is a need for this multitude of overlapped bands to be disentangled. Using deuterium for substitution, large frequency shifts and intensity changes are observed. When one can prepare selectively deuterated analog polymers to isolate the spectral contributions of each type of C—H for example, the data processing capabilities of FT-IR allow considerable help in making the proper band assignments [216]. For example the three deuterated analog polymers of poly(methyl methacrylate) (PMMA) have been prepared, that is with only the methyl groups undeuterated, the ester methyl groups undeuterated, and only the methylene groups undeuterated [216]. In this fashion each of the carbon-hydrogen functional groups gives rise to the spectrum characteristic of that group without band overlap of the other carbon-hydrogen motions. The spectral assignments are particularly simplified and the expected band overlap is completely eliminated — at least to the extent that the deuteration is complete. The spectra are normalized relative to each other with elimination of the undeuterated portions. Then the digital spectra of PMMA can be synthesized by

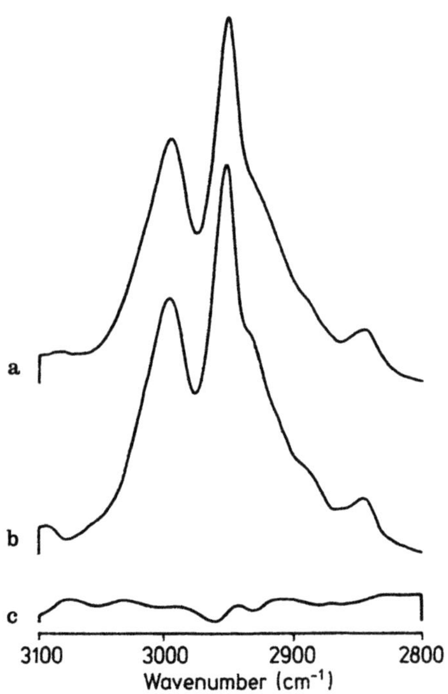

Fig. 18a—c. Infrared absorption spectra of PMMA in the C—H stretching vibrations region. **a**) experimental PMMA spectrum; **b**) digital PMMA spectrum: coaddition of PMMA—CD_2 —CD_3 = PMMA—CD_2—OCD_3 + PMMA—CD_3—OCD_3 (1:1:1) using the 1732 cm^3 — 1 absorption band [ν(C=0] for normalization; **c**) difference spectrum: experimental PPMA — digital PMMA using 1732 cm^{-1} absorption band [ν(C=0)] for compensation.
Ref. Fig. 4 from Stoil K. Dirlikov and Jack L. Koenig Appl. Spectr., Vol. *33*, No. 6, 1979, pp. 555–561

adding the spectra of the analog polymers. In this manner the assignment of the 14 absorption bands in the stretching vibrations region of PMMA were made. Figure 18 shows the observed experimental spectrum of PMMA and the spectra obtained by coaddition of the spectra of the deuterated analogs as well as the difference spectrum which illustrates that the agreement is excellent.

Similarly other isotopic species can be used but the magnitude of the effects produced is smaller. With FT-IR, these small differences can be observed easily using difference spectra and bands associated with the oxygen species have been identified in PMMA used oxygen-18 substitution [217].

7 Studies of Polymer Chemistry Using FT-IR

7.1 Description of Method

FT-IR has a number of advantages for studying chemical reactions involving polymers. Compared to other analytical techniques such as NMR, the FT-IR sensitivity is very high and has the ability to study a wide range of different sample types under a variety of environmental conditions. If the lifetimes of intermediate species are of the order of 3 sec, the complete spectrum can be obtained using FT-IR. In addition, the kinetics of the reactions can also be studied as will be indicated in a later section.

For example, for studying polymer degradation, the problem with infrared measurements using dispersive instruments is the insensitivity, making it difficult to determine the initial point of attack on the molecule. After extensive degradation the spectra exhibit only broad, poorly resolved bands as a consequence of the overlap of polymer bands arising from the diverse products of oxidative attack. Using absorbance subtraction, small differences in the nature of the reacting or product species can be detected and under appropriate circumstances quantitatively measured. A differential series of spectra can be obtained plotting each spectrum against a reference spectrum. The reference spectrum can be the initial spectrum or the immediately preceding spectrum. The differences in the response of the different infrared bands will reflect the relative reactivities of the chemical groups. This advantage allows the study of complex mechanisms of degradation by FT-IR.

7.2 Oxidation of Polymers

Some of the early work using FT-IR involved the study of polymer degradation processes [218,219,220,221,222]. The FT-IR measurements were particularly useful in providing information concerning the initial product formation which had previously eluded detection. Since one can study exactly the same portion of the sample, spectral subtraction are made on a one-to-one basis at the early stages of the reaction and the resultant difference spectra can be magnified (as high as 100–200 times) to bring out small spectral features. Thus, in the difference spectrum, any increases in absorbances above the baselines are due to increases in absorbing species (oxidation products) and any decreases are due to losses of absorbing species (point

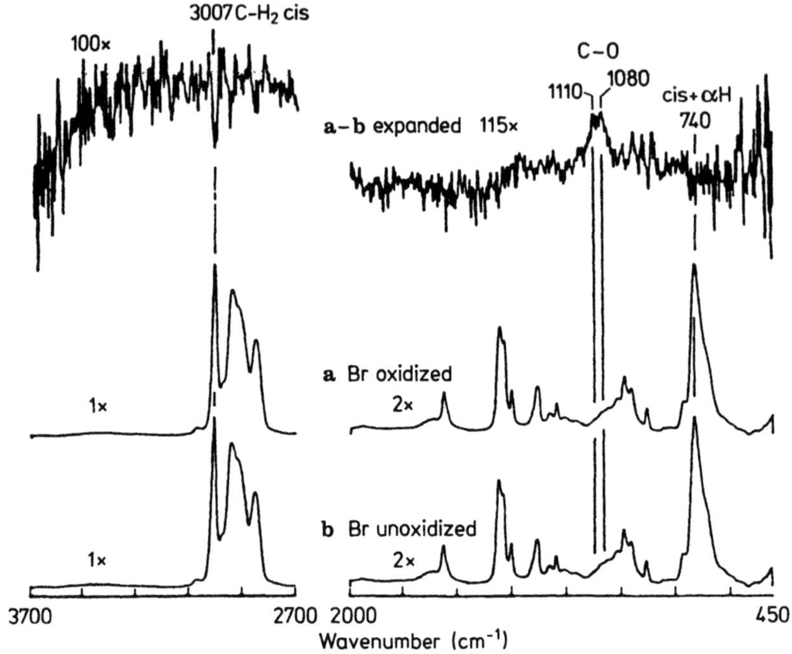

Fig. 19a and b. Cis-1,4-polybutadiene oxidation, 1 h at 25 °C. Bottom, unoxidized; center, oxidized; top, difference spectrum.
Ref.: Fig. 1 from R. L. Pecsok, P. C. Painter, J. R. Shelton and J. L. Koenig, Rubber Chem. and Technol., Vol. 49, 110 1976

of attack on the polymer chain). (Fig. 19) shows the spectra of unoxidized cis-1,4-polybutadiene rubber and the same sample oxidized for 1 hour at 25 °C in air. Although no differences are observed visually, the difference spectrum (magnified 100 times) shows the appearance of a number of spectral differences [220]. The initial point of attack is indicated by the loss of cis-methine groups as shown by the negative 3007 and 740 cm^{-1} bands. The lack of hydroxyl absorption at this early stage and the initial presence of C—O groups at 1080–1110 cm^{-1} suggests the initial formation of cyclic alkoxyperoxides. In fact the presence of the two absorption bands suggest the appearance of two different types of alkoxyperoxides and a mechanism has been proposed which is consistent with this observation [220]. Subsequent studies were made of the inhibited oxidation of cis-1,4-polybutadiene with three different antioxidants [223] of the phenolic and arylamine types. The initial observation in this work was the inability to effectivelly carry out spectral subtractions because none of the butadiene bands could be relied upon as a proper internal mass standard since they all change in band shape and intensity with the addition of the antioxidants. These types of spectral changes result from the presence of specific chemical interactions of the antioxidants with the rubber [223]. The spectral results give some insight into the stabilization mechanisms of these antioxidants [223]. These degradation reactions are complex including a series of simultaneous reactions involving bond scission through oxidative or thermal cracking, depolymerization, chain crosslinking through the formation of C—C bridges during radical recombination, and mutual reactions of the oxidative products.

The oxidation of other rubbers has been studied by FT-IR including polychloroprenes [224]. These results suggest that the thermal oxidation of polychloroprenes involves the 1,2 and 3,4-structural irregularities in the initial stage. In particular, it is felt that the initial step is the abstraction of a tertiary allylic chlorine or hydrogen from the 1,2 or 3,4 units yielding a tertiary carbon radical.

A series of papers have reported FT-IR studies of the thermal degradation of polyacrylonitrile [225, 226, 227, 228, 229, 230]. These FT-IR measurements indicate that the nitrile groups play an important role in the degradation process even in the initial stages and show a steady decrease in nitrile with time both in air and under reduced pressure. These results are contrary to previous dispersive infrared measurements

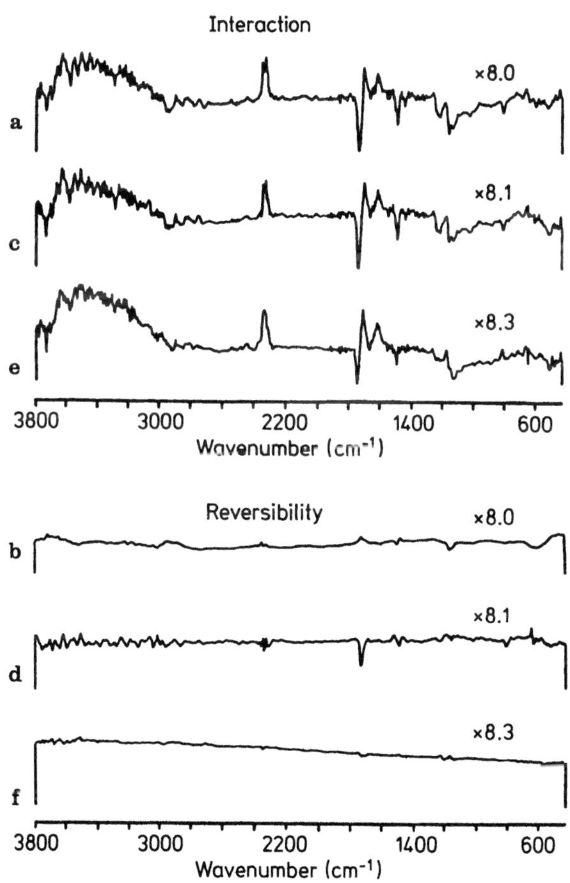

Fig. 20a—f. Reversibility of the epoxy/water interaction through three cycles of water sorption and redrying: **a**), **c**), **e**) interaction spectra created by subtracting the original dry epoxy film spectrum from the epoxy/sorbed moisture spectrum; **b**), **d**), **f**) difference spectra of the redried epoxy film minus the spectrum of the original dry epoxy film spectrum. Spectra are lettered in the order in which they were obtained. Ref. Fig. 4 from M. K. Antoon and J. L. Koenig, J. of Polym. Sci.: Polym. Phys. Ed., Vol. 19, 1567–1575 (1981)

which lacked suitable sensitivity. Degradation of the deuterated derivative confirmed the imine-enamine tautomerism followed by oxidation leading to the formation of pyridone structures [226]. The introduction of a small amount of acrylic comonomer in the PAN chain increases the rate of degradation [227] and each monomer accelerates the degradation in a specific fashion which depends on its chemical structure. In spite of the relatively low concentration (about 4%) the FT-IR data yield significant information concerning the structures formed and the mechanism of these complex degradation systems.

Oxidation studies have been made for cured epoxy resins and the relative stability of the functional groups was established by following the changes in the absorbance ratios of bands associated with the particular functional group [231]. Hence the most unstable groups can be determined easily and their reactions separated from the more stable units. The irreversible [232] and reversible [233] effects of moisture on epoxy resin systems have also been studied by FT-IR using the difference spectrum method [234]. (Fig. 20) shows the interaction spectra obtained after three cycles of water sorption and redrying. The reversible nature of the interactions are clearly demonstrated.

7.3 Irradiation Damage of Polymers

The effects of irradiation damage on polymers has also been followed using FT-IR [219, 235, 236], by obtaining difference spectra after specific exposure durations in a variety of environments. For polyethylene [219], the difference spectra revealed an increase in the vinyl end groups and an increase of ketonic carbonyl and trans-vinylidene double bonds. It was found that for every free radical formed approximately 10 carbonyl groups and 2 carbon-carbon double bonds were formed [236].

Irradiation of polytetrafluoroethylene produced acid end groups and the scission occurs in both the crystalline and amorphous regions. When a sample of polytetrafluoroethylene is used with a high initial crystallinity, irradiation decreases the crystallinity [235].

7.4 Mechanical Reversion in Polymers

One of the problems in the utilization and processing of rubber is the reversion process. Reversion occurs when continued curing causes the polymers to exhibit a loss of physical and mechanical properties such as tensile strength, stiffness, resilience and wear resistance. The nature of the chemical changes occuring during this reversion process have been studied for unfilled [242] and carbon black filled natural rubber [243]. It was found that a correlation exists between the amount of trans-methine structure in the vulcanizate. The presence of the trans-methine structure was detected by measuring the absorbance of the 965 cm^{-1} band using FT-IR. Initially, no trans-methine structure is observed but is detectable after 4–5 minutes of curing with an accelerator. A vulcanization process leading to the formation of this trans-methine structure has been proposed [242]. Carbon-black filled systems have a better reversion resistance due to the restriction of the double-bond migration in the polymer chains and less trans-methine formation [243, 244].

8 Polymer Structure Analysis Using FT-IR

The potential applications of FT-IR to the determination of polymer structure are many. In a number of areas, the impact of FT-IR has been very significant including the structure of polymer blends, polymer surfaces and the structural changes induced by mechanical deformation. These topics will be discussed in detail below.

8.1 Study of Polymer Blends

Although polymer blends are difficult to study by other techniques, blends represent a particularly good example of the utility of the data processing capability for elucidation of structural information. The approach is quite simple [247]. If the two polymers are completely incompatible, the infrared spectrum of the blend should be the sum of the spectra of the two pure components. Incompatiblity implies phase separation and the polymers are presumed to be in a state similar to the pure component and do not recognize the presence of the other component. If the polymers are compatible, there is only one phase and the individual polymer chains are "solubilized" to the extent that there is a distinct chemical interaction between the two different polymers. This interaction leads to considerable differences between the spectrum of the polymer in the blend and as a pure component. This spectroscopic difference can easily be detected by subtracting out the contributions of both homopolymers and the isolation of the "interaction" spectrum. These ideas were tested using factor analysis for incompatible and compatible blends of 2,6-dimethyl poly(phenylene oxide) PPO and polystyrene (PS) [100]. The incompatible blend had the polystyrene chlorinated in the para position while the compatible system was unchlorinated. The incompatible system exhibited only two components i.e. the spectra of the homopolymers while the compatible blend showed three components, i.e., the spectra of the two homopolymers and the interaction spectrum arising from spectral differences resulting from the chemical interactions in the compatible blend. The interaction spectrum can be isolated and used to quantitatively characterize the amount of interaction for blends of different proportions prepared by different processes.

Coleman et al. [247] reported the spectra of different proportions of poly(vinylidene fluoride) PVDF and atactic poly(methyl methacrylate) PMMA. At a level of 75/25 PVDF/PMMA the blend is incompatible and the spectra of the blend can be synthesized by addition of the spectra of the pure components in the appropriate amounts. On the other hand, a blend composition of 39:61 had an infrared spectrum which could not be approximated by absorbance addition of the two pure spectra. A carbonyl band at 1718 cm^{-1} was observed and indicates a distinct interaction involving the carbonyl groups. The spectra of the PVDF shows that a conformational change has been induced in the compatible blend but only a fraction of the PVDF is involved in the conformational change. Allara [249, 250, 251] cautioned that some of these spectroscopic effects in polymer blends may arise from dispersion effects in the difference spectra rather than chemical effects. Refractive index differences between the pure component and the blend can alter the band shapes and lead to frequency shifts to lower frequencies and in general the frequency shifts are to lower frequencies.

There is no doubt that this dispersion effect occurs but the magnitude of the spectral differences appear in most cases to be considerably larger than would be predicted by dispersion effects. For example, the poly(ε-caprolactone) (PCL) and poly (vinyl chloride) (PVC) blend has been studied [252,253] and for this system the refractive indices are identical. In this case, there are obvious frequency shifts and broadening of the carbonyl band as a function of PVC concentration as shown in (Fig. 21). Nine percent of the original area of the carbonyl peak is involved in the shifted frequency absorption. Clearly, for this system, chemical interaction effects are being observed. In fact, PCL can be viewed as a macromolecular plasticizer for PVC. The blend system poly(β-propiolactone) PPL and PVC was studied [253]. In contrast to the PCL/PVC system, the PPL/PVC system was incompatible over the entire range of compositions.

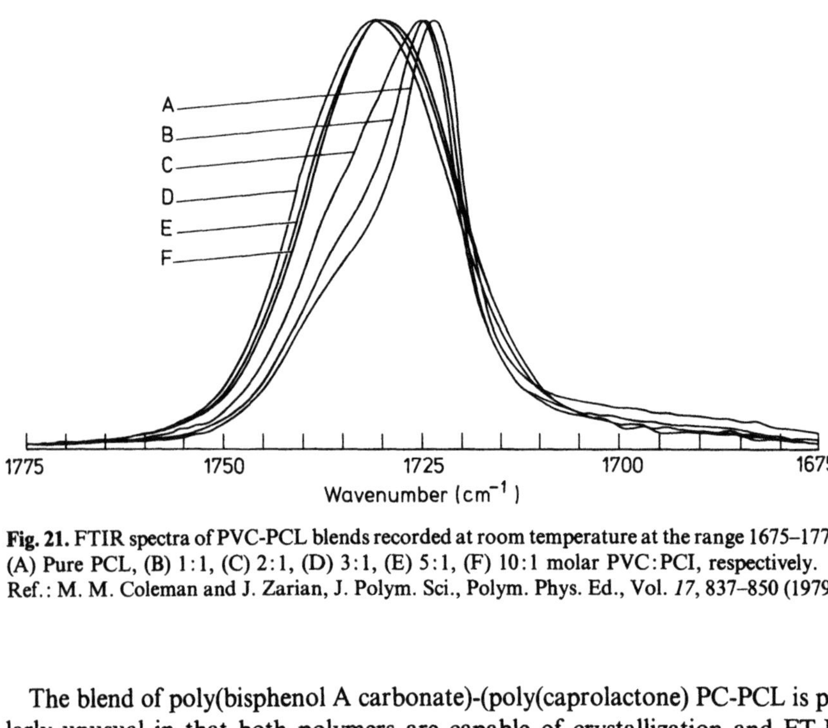

Fig. 21. FTIR spectra of PVC-PCL blends recorded at room temperature at the range 1675–1775 cm^{-1}. (A) Pure PCL, (B) 1:1, (C) 2:1, (D) 3:1, (E) 5:1, (F) 10:1 molar PVC:PCI, respectively. Ref.: M. M. Coleman and J. Zarian, J. Polym. Sci., Polym. Phys. Ed., Vol. *17*, 837–850 (1979)

The blend of poly(bisphenol A carbonate)-(poly(caprolactone) PC-PCL is particularly unusual in that both polymers are capable of crystallization and FT-IR has been used to study the state of order in these blends as a function of the method of preparation [254,255]. In this case, PCL is a macromolecular plasticizer for PC. The PCL becomes progressively less crystalline as the concentration of PC increases. PC is amorphous if the blend is cast from methylene chloride but semicrystalline if cast from tetrahydrofuran. When PC in the pure form is exposed to acetone, it will not crystallize, but in the blend, exposure of acetone causes the PC to crystallize which emphasizes the additional mobility of the PC in the blend.

The orientational behavior of the polymer chains in blends has also been studied [256]. It was observed that the PS chains increased regularly with concentration up to

24% PPO and then remains constant while the PPO chains does not exhibit orientational behavior which depends on the concentration. These results are explained in terms of the different relaxation behavior of the two types of chains.

8.2 Polymer Surfaces and Interfaces

In spite of the development of the reflection methods indicated above, the study of surfaces using transmission measurements remains the most used technique particularly for inorganic and organic substrates which can be prepared in a form appropriate for light transmission. Of course, metal and other nontransmitting surfaces cannot be used in the transmission mode. It was recognized when the first absorbance subtraction results for polymer surface studies were reported [7], that transmission spectroscopy could be used to study surface species. If one could obtain by infrared transmission the spectra of a bulk sample and that of the same bulk sample which had been given a surface treatment, spectral subtraction could be used to remove the interference due to the bulk phase and the resultant difference spectrum would reflect the differences in the two surfaces before and after treatment. If the untreated sample were subtracted from the treated sample, positive absorbance bands would arise from a higher concentration of surface species in the treated sample and negative bands from species preferentially found in the untreated surfaces. In this manner, one can determine the nature of the chemical reactions occurring with surface treatments. Since the initial observations, infrared transmission measurements using absorbance subtraction have become quite popular. This interest arises from the high sensitivity of the transmission measurements, the quantitative potential of the measurements, and the ease of the experiments when the samples can be prepared in the appropriate format. The sample preparation techniques must not affect the surface in a detectable manner and for many surfaces, this limitation is a severe experimental problem. If the substrates to be examined can be initially prepared as thin films the transmission technique is superior to any of the reflection techniques outlined above. For example for samples like polymer films, the technique is ideal since the spectrum of the original film can be directly compared with the spectrum of the film after treatment. The difference spectrum can be scale expanded to enhance the sensitivity and the surface species can be quantitatively measured [7].

Surface studies have been made on substrates which are highly absorbing like glass [257,258]. Silane coupling agents are used to enhance adhesion of polymers to glass surfaces, but the nature, structure, amount, orientation, and organization has not been determined. For high surface area silica, monolayer coverage could be detected and the presence of a polysiloxane structure observed [257]. Additionally, a chemical reaction leading to the formation of interfacial bonds was found. Similiar studies on E-glass fibers revealed multilayer formation (number of layers depending on the concentration of the treatment solution) [258]. But in this case, where the surface area was substantially lower, the number of interfacial bonds was insufficient to be detected [258]. The chemical reactions between the organic portion of the silane coupling agent and the polymer matrix can be observed [259] as copolymerization between gamma-methacryloxypropyl silane and styrene monomer occurs at the treated glass interface. The silane coupling agent interface has a degree of order in the molecular organization on the surface and the extent of order depends on the nature of the coupling agent and the treatment conditions including concentration and temperature of the treatment solution. When a cyclohexyl functional silane is used, crystalline layers of silanetriol on the glass fibers are observed. Vinyl and aminopropyltriethoxysilanes show less order [260] on the glass surface. The effect of hydrolysis and drying on the molecular structure of the interphase can also be investigated as well as the hydrolytic stability [261, 262]. A summary of these results has been reported [263].

Aminofunctional silane coupling agents show unique solution properties when nitrogen is on the gamma-carbon atom. These alkoxysilanes are hydrolyzed almost immediately in water, and form a dilute aqueous solution of good stability. This stability has been proposed to arise through zwitterion formation with one of the unreacted silanols or as a five-membered ring with the nitrogen atom interacting with the silicon atom. FT-IR has been used to study these structures and their reactions [264]. Similar studies have been reported for multiple amino silane coupling agents [265]. The nature of the interfacial reactions between the amino silanes and the epoxy matrix resins have received attention [266].

As a result of a series of papers, the nature of the silane coupling agent interphase has been ascertained [267, 268, 269, 270, 271, 272, 273, 274, 275, 276, 277]. There are three layers consisting of different molecular structures as one proceeds from the glass surface. First, there is a well-ordered highly crosslinked layer of between 1 and 50 molecules in thickness. This initial layer is chemically bonded to the glass surface. The middle portion of the interphase is less ordered and less highly crosslinked and can chemically react with the matrix resin to form interfacial bonds. The outer layer is a physisorbed layer which is very loosely bonded and is mechanically weak. This molecular model is shown in Fig. 22. Experiments based on this model have lead to improved initial strengths and service lifetimes of composites.

The adhesion of rubber to PET fibers has been studied using FT-IR [278, 279]. The kinetics of the chemical reactions occurring with the coupling agent and the end groups of the PET fiber have been measured and the results extrapolated to a quantitative determination of the adhesive bonding during fabrication [278, 279].

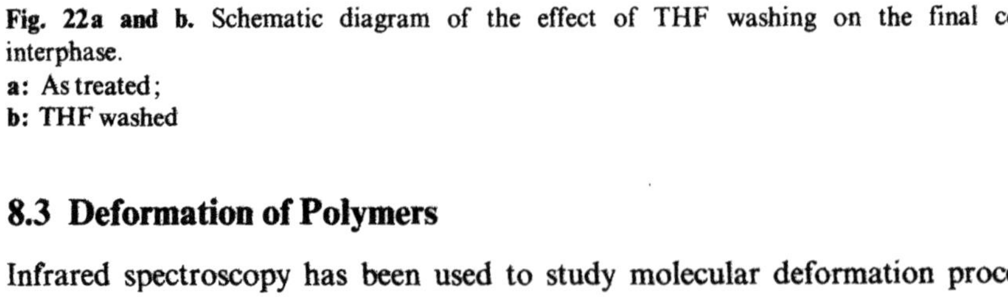

Fig. 22a and b. Schematic diagram of the effect of THF washing on the final composite interphase.
a: As treated;
b: THF washed

8.3 Deformation of Polymers

Infrared spectroscopy has been used to study molecular deformation processes in polymers as a function of stress, strain, time and temperature. Wool has summarized the spectral effects resulting from deformation [280]. The infrared spectra show vibrational frequency shifts and intensity variations as a result of macroscopic deformation. The frequency shifts can arise from stress introducing structural changes, molecular orientation, conformational differences, chain fracture leading to new bands, and molecular strain. However, although the spectroscopic changes resulting from deformation are easily observed, the interpretations in terms of structural and mechanistic differences are controversial. An analysis of the various studies, to date, reveal a number of interpretations. Some investigators have utilized a bond strain effect which results in frequency and intensity shifts [280, 281]. Other investigators have used conformation transitions induced by stress to interpret the results [282, 283, 284]. Finally, the frequency shifts have been explained by a reduction of the force constant as a result of a disruption of intermolecular interaction in the stressed polymer [285]. From these results, it is obvious that each polymer will show different spectral features depending on its microstructure. A polymer like PET can exhibit all of the above structural changes and therefore the corresponding spectroscopic changes [282]. This polymer has been studied extensively and the band assignments are considered to be satisfactory making the interpretation of the effects of stress easier. The initial stress of deformation is borne by the interfibrillar (amorphous) regions. The stress increases the crystallinity.

Free radicals are also produced by chain scission during deformation of polyethylene and FT-IR has been used to follow this process [237]. The polyethylene samples were unaxially drawn and the resultant spectra corrected for orientation. An increase in the vinyl and methyl end groups created by decay of the free radicals occured in going from draw ratios of 5 to 20 [44]. A similar study involving deformation was made of polystyrene [246] and a comparison demonstrated between the results of thermal and mechanical degradation [245].

Mechanical deformation induces orientation into the polymer samples and polarized infrared can be used to characterize this orientation either by direct measurement of the dichroic ratio [287, 289] or by spectral subtraction [286], three dimensional sample tilting [68, 286], or internal reflection spectroscopy [130].

The orientation induced by drawing has also been followed for partially oriented PET fibers. In this case where unaxial orientation can be presumed, it is possible to calculate the relative amounts of the trans and gauche isomers. The measurements indicate that increased orientation of the amorphous phase leads to an increase in the trans isomer [288].

9 Time-Dependent Phenomena in Polymers

The Fellgett and Jacquinot advantages manifest themselves by allowing one to obtain the complete infrared spectrum in the order of a second, so short-time phenomena (1s $< t <$ 600s) can be studied [16]. The measuring time for a dispersive instrument is two orders of magnitude larger. This improvement in the speed of measurement allows the monitoring of short-time spectroscopic changes for a variety of processes in polymers, including kinetics [13, 44], crystallization [290], heating [291], orientation [292] and relaxation [292]. A rather detailed presentation of the effects, using the rapid scan speed, for polymer studies exists [17] so only a general discussion with some of the new results will be presented here.

In a typical kinetics experiment, one obtains between 100 and 2500 spectra so there is a need for the storage and processing of the data. To store a single spectrum at 2 cm^{-1} resolution requires 4576 words. A disk cartridge can store 2,286,592 words or 433 spectra with a resolution of 2 cm^{-1} so such disks supply sufficient storage capacity. It is also necessary to program the computer to process the data since manipulation of the data by inspection would require too much time. With suitable programs, the required data can be reduced in a matter of seconds after acquisition. The plot of the absorbance of a characteristic group frequency is a kinetic profile of the process. It is also possible to use differential spectra by ratioing or subtracting each spectrum from the preceding one with respect to time [293].

Another requirement for kinetic studies is an appropriate interface between the spectrophotometer and the sampling device so that one can clock the data aquisition and activate the instrument at suitable times. For the studies to be described, suitable interfaces have been constructed [16].

9.1 Studies of Curing of Polymers

The kinetics of the curing process for multifunctional monomers can be carried out using FT-IR in the manner described above [44, 238, 239]. In particular the "time-lapse" type of measurements can be made where the spectra are recorded at predefined intervals depending on the speed of the reaction [240]. The tertiary amine-catalyzed copolymerization of cyclic anhydrides and epoxy resins has been studied using absorbance subtraction from the early stages through gelation [44]. The kinetics of the polymerization were analyzed in terms of the current mechanism and it was found that modifications are necessary particularly in the role of the hydroxyl groups which were found to catalyze the curing process. It was also revealed that the structures produced during the early stages of the curing were different from those formed during the final stages. Hence, the nature of the chemical reaction has changed. Similar studies were made using 1:phenoxy-2:propanol as a catalyst [241].

9.2 Crystallization of Polymers

The kinetics of the crystallization process can be followed using the FT-IR technique [290]. A particularly interesting example comes from the study of the polymer blends of PVDF/PMMA where the crystallization of the alpha and beta forms have been followed during heating of the blend samples which had been quenched from the melt and crystallized by heating at 2 °K/min in the spectrometer. When the blend has 70 wt% PVDF the beta crystal form is obtained directly but for other compositions the alpha form is dominant or unique.

9.3 Heating Effects in Polymers

One of the methods of following changes in intermolecular and intramolecular effects is to heat the sample and compare the changes in the spectra as the sample is being heated. For example between 448 °K and 478 °K, it has been shown by differential scanning calorimetry that the hard segments melt in polyesteraurethanes. The spectroscopic changes occurring during the heating (5 °K/min) of the polyesterurethane film have been followed [291]. The intensity changes of the bands associated with free and hydrogen bonded N—H groups indicate a shift in the equilibrium concentration of hydrogen-bonded and non-hydrogen bonded NH groups. The effect of heating on the changes in conformation of poly(tetramethylene terephthalate) has also been studied during the heating process [67]. The sample was studied during heating from 338 °K to 468 °K and scanned in time intervals of 20 s. The spectrum shows dramatic changes in the band at 910 cm^{-1} while the other bands in the spectrum show considerably less change. The 910 cm^{-1} band is associated with the alpha-conformation.

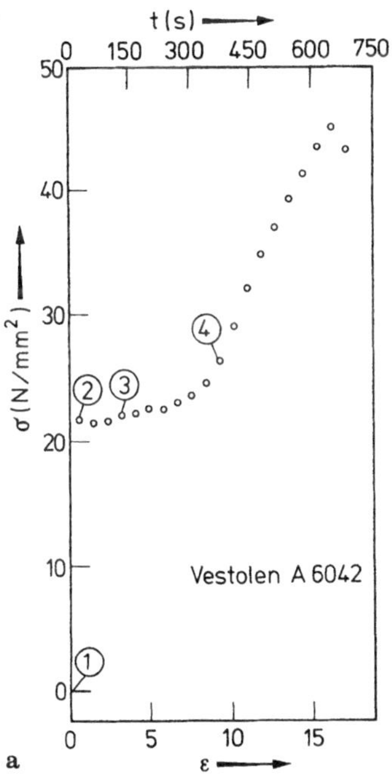

a

Fig. 23a. Stress-strain-diagram of a Polyethylene (Vestolen A 6042) film (stretching velocity: 0,26 mm/s); **b**) Experimental (row I), synthesized (row II), and resolved (row III) bands of the CH_2-rocking bands. The experimental spectra were scanned at the indicated positions (circled numbers) of the stress-strain-diagram (**a**).

Ref.: Fig. 2 from K. Holland-Moritz, and K. van Werden, Makromol. Chem., *182*, 651–655 (1981)

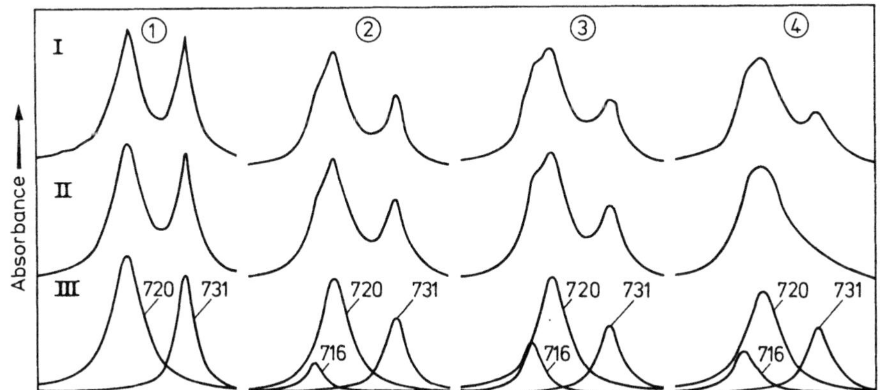

b

9.4 Dynamic Deformation or fatigue of Polymers

With the speed of the instrument it is possible to obtain the infrared spectrum during deformation [295]. Recently, the spectral behavior of the doublets at 1473/1463 cm^{-1} and 730/720 cm^{-1} have been followed during Polyethylene deformation. Fig. 23 shows the spectra obtained and the corresponding point on the stress-strain curve. For small strains, the orientation increases only slightly. When the neck forms, the spectra reflect a small rotation of the b-axes into the stretching direction and an opposite movement of a-axes. During the stretching process the significant shoulder at 716 cm^{-1} appears. The dichroic ratios have also been obtained for the rocking modes and are shown in (Fig. 24). Isotactic polypropylene has been studied in a similar fashion and the spectra were recorded at a 13.5% strain interval, while the sample was uniaxially drawn at 302 °K with an elongation rate of 67% per minute. Large parallel dichroisms were observed in the region from 50 to about 150% strain indicating a preferential parallel alignment of the polymer helix axes with the direction of stretch [297].

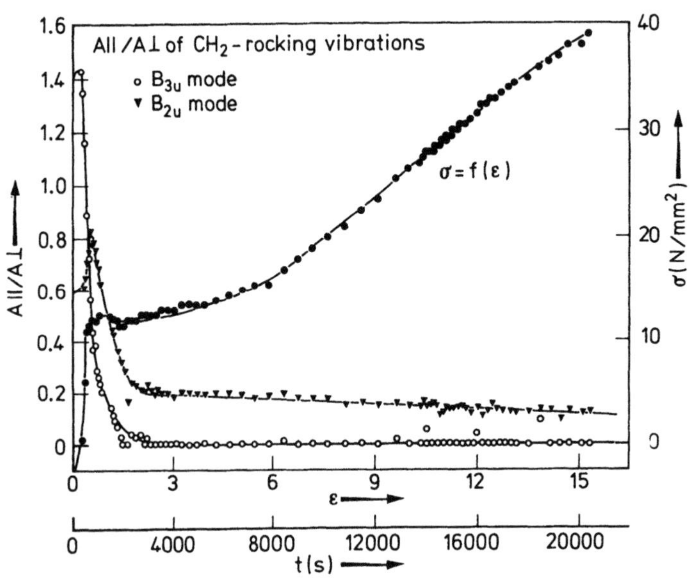

Fig. 24. Dichroic ratios (A‖/A⊥) and stress (σ) as function of strain (ε) (stretching velocity: 0.008 mm/s. x: CH_2-rocking band ($B_{3\mu}$) between 736 and 726 cm^{-1}, ∇: CH_2-rocking band ($B_{2\mu}$) between 726 and 710 cm^{-1}, + : stress-strain-diagram.
Ref.: Fig. 4 from K. Holland-Moritz, I. Holland-Moritz and K. van Werden, Colloid & Polym., Sci., *259*, 156–162 (1981)

Perhaps, the most dramatic example of the effects of drawing on the spectra of a polymer comes from the stress-induced crystal transition of poly(butylene terephthalate) [292, 294, 298, 299]. The phase transition of PTMT has been investigated above and below the glass transition temperature by means of dynamic infrared measurements

[298]. The results indicate that the temperature of the sample being drawn plays a significant role in the nature of the transformation. The phase transition is reversible and occurs from one triclinic unit cell to another triclinic cell [294]. Both forms exist and are separated by only a small potential energy barrier. This barrier can be surmounted by using the energy of the applied stress. Since the transformation is reversible there must be two opposing forces. One force is the packing of the terephthalic acid groups and the opposing force is the lowest-energy conformation of the tetramethylene chain (as seen in the beta phase). In this polymer, the stress relaxation can also be followed [292]. The spectral data suggest that the stress relaxation mechanism occurs primarily by a slippage process of the chain segments while retaining their conformational character.

9.5 Time-Resolved Spectroscopy

When one is investigating time-dependent repetitive processes such as fatigue measurements in polymers, it is possible to do stroboscopic measurements in such a fashion as to obtain the complete spectrum of the sample in the time domain of microseconds [300, 301]. The technique is based on the existence, at any given time, of a small interval of time during which the event appears stationary. During this time interval, the interferogram signal, at a sampling position corresponding to a path difference of x_n is obtained. With a suitable offset in the timing of the collection of data during the next cycle, the data at a point $x + 1$ can be obtained for the interferogram at time t_n. By assembling the recorded signals having a common t, say t_n, these complete interferogram signals from zero to the maximum retardation are obtained and can be transformed to give a spectrum at time t_n. In fact, n interferograms are obtained each corresponding to a specific time. If the process is repeated a sufficient number of times, the timing is accurate and the sampling of the data properly sorted, one can obtain a spectrum after Fourier transformation with a time resolution of microseconds. For an experiment involving an oscillatory strain, one can obtain the spectrum at the initial time with no stress or strain, and at intervals of time

$$t_n = t_0 + (x/v)$$

corresponding to different strain levels as the sample is deformed. In the above equation t_n is the nth data point, t_0 is a constant time interval at the start of the experiment, x is the distance of the mirror drive and v is the velocity of the mirror. The timing or distance of mirror travel is obtained from the zero crossings of the interferogram of the internal HeNe laser and the data can be collected systematically based on a chosen number of zero crossings. With a suitable offset and careful time sorting of the data, the complete interferogram can be obtained [302]. The mechanics of the technique are shown in Fig. 25 which shows the timing of the data collection by the times, t_1, t_2, etc., and the A/D conversion at that time. The offset indicates the method of collecting the complete interferogram, which for the example is limited to only five data files. It is crucial to the success of the experiment that the sample does not

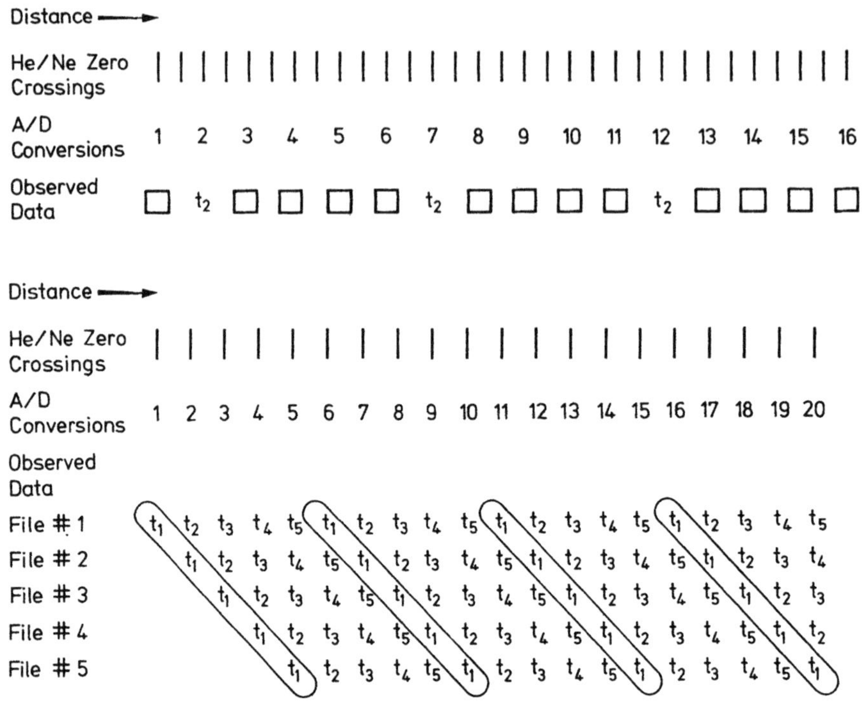

Fig. 25. Diagramm of time-sorting experiment.
Ref.: Fig. 4 from W. G. Fateley, and J. L. Koenig, J. of Polym. Sci., Polym. Lett. Ed., *20*, 445–452 (1982)

undergo irreversible changes during the cyclic deformation for the time period of the data collection or the data will not be representative of the portion of the cyclic deformation process. Much of the early work in time-resolved infrared spectroscopy with FT-IR instruments suffered from a time drift of the sample during the accumulation period [303]. Since polymers can undergo thousands of cycles in fatigue if the strain level is not high, it appears that the fatigue experiment in polymers represents a nearly ideal problem for investigation using time-resolved spectroscopy (TRS).

The first work reported in this area was an investigation of isotactic polypropylene which was fatigued at a rate of 10 hz with an elongation of one to five percent. The spectra showed no shifts in frequency but reversible intensity changes were observed. The data were interpreted in terms of the presence of a "smectic-like" structure as a result of the cyclic stress [304,305].

The response of ion-containing ethylene-methacrylic acid copolymers has also been investigated [306]. The strain amplitude of the samples was 2% and the period of external stress was 50,000 microseconds. Changes in the dichroic ratio of the 2673 cm^{-1} band were observed as a function of strain in times as short as 200 microseconds. Other spectroscopic changes were observed but not interpreted in terms of structural changes.

10 Temperature Effects on Spectra of Polymers

Polymers in end use are subjected to a range of temperatures and a knowledge of their structure and performance under different thermal conditions is required. FT-IR offers a method of studying some of the effects of temperature although the problem has been complicated in the past, not by experimental limitations, but rather by theoretical problems of interpretations of the spectral effects. The problems arise in attempting to separate out the spectral changes arising from temperature-induced structural variations and temperature-induced spectroscopic changes in frequency and extinction coefficients. However, recently it has been possible to separate these two effects using the absorbance ratio method to calculate the spectra of the contributing structures at the measurement temperature [307].

First, let us consider infrared studies at cryogenic temperatures as a probe of transitions in polymers. The temperature of the extinction coefficient is usually assumed to have a linear dependence [308].

$$\varepsilon = \varepsilon(T_r) + a(T - T_r)$$

where a is usually negative and T_r is a reference temperature. This type of equation arises since the intermolecular expansion reduces the induced dipole moment of the interaction and, therefore, the intensity. The effect is most pronounced for vibrational groups which are highly polar. When the absorptions are related to the various components of the structures, then infrared can be used as a probe of the thermal responses of these structures. A change in slope in an intensity versus temperature plot indicates the existence of a transition. Koenig and coworkers have used this rationale to interpret changes in the intensities at cryogenic temperatures of PET [309], and PS [310] using dispersive infrared measurements. No concentration changes are expected for these low temperature regions and this infrared type of molecular dilatometry should observe the same transitions as bulk thermal expansion measurements. An additional benefit of the infrared studies is an indication of the molecular groups involved in the transition. More recently, FT-IR measurements at low temperatures have been made on polyethylene [311]. The higher signal-to-noise and data processing capability allow detection of smaller effects with the FT-IR. Hence, it was possible establish the existence of a dual glass transition polyethylene [311]. Figure 26 shows the normalized absorbance of the 731 cm^{-1} band as a function of temperature for three different samples of polyethylene. The error in measurement of the absorbance is 3%. The intensity of the slow crystallized sample shows the lower glass transition by a change in slope at 190 ± 5 °K. The major change in intensity with temperature occurs at 240 ± 5 °K for the rapidly quenched sample while for the annealed sample only minor inflections are observed.

FT-IR has also been used to study the molecular mechanisms of transitions in atactic polystyrene above the glass transition [312]. These infrared measurements detected two transitions above the glass transition. The intensity data was tested for curvature by computer analysis using a least squares fitting program. An error analysis was used to select between the use of linear segments to fit the data compared to a continuous curvature fit. The lower transition (165 C) is molecular weight independent while the higher transition (200 C) has the same molecular weight dependence as the

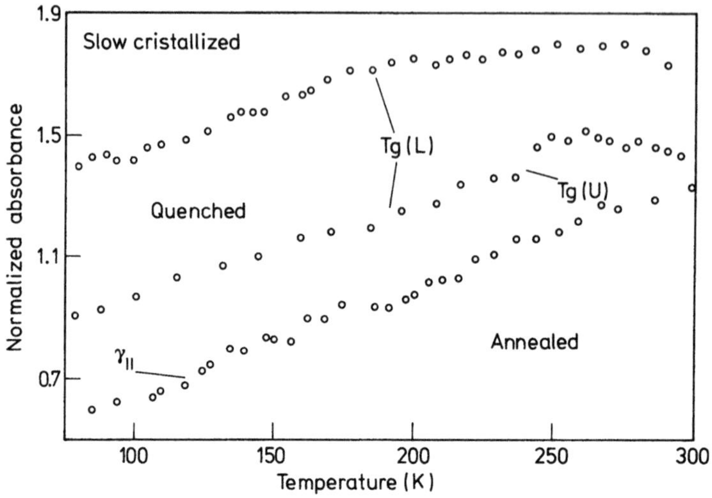

Fig. 26. Normalized interpolated peak height of 731 cm^{-1} absorption vs. temperature. Maximum error of measurement is 3% (within data marker)

glass transition. Apparently, the higher transition involves a conformational effect while the lower transition arises from long range motions [312].

Ordinarily, changes in the temperature cause a change in the distribution of various conformations or structures present in the system; then the infrared spectrum will reflect these changes. Anton [313] examined a number of polymer systems using temperature variation. He studied the glass transition of a number of polymers in this fashion. The ratios of bonded and free hydrogen bonds in nylons and the extent of dimerization in methacrylic acid polymers have also been studied as a function of temperature [314, 315, 316]. FT-IR has been used to quantitatively study the extent of hydrogen bonding and its temperature dependence in a segmented polyurethane elastomer synthesized from 2,6-toluene diisocyanate, 1,4 butanediol, and poly-(tetramethylene oxide) [317]. In this polyurethane, there are two very strong carbonyl stretching modes. The band at 1733 cm^{-1} is assigned to the free carbonyl while the one at 1691 cm^{-1} is assigned to the hydrogen bonded carbonyl stretching mode. The free N—H stretching vibration is a shoulder on the high-frequency side of the hydrogen bonded N—H present at 3283 cm^{-1}. The hydrogen bonded N—H peak absorbance decreases at the higher temperatures with a corresponding increase in the free N—H peak absorbance. The hydrogen bonded N—H band maximum also shifts from 3283 cm^{-1} to 3290 cm^{-1}. At 29 °C, it was found that 88% of the N—H groups are hydrogen bonded and at 151 °C only 50% were bonded. The equilibrium constant for the dissociation of the hydrogen bonded N—H was determined as a function of temperature. The enthalpy and entropy of hydrogen bond dissociation in this polyurethane were found to be 29 KJ/mol and 64 Jmol/K, respectively, based on the N—H concentration. A similar study was performed for the polyurethane system with the hard segments based on p,p-diphenylmethane diisocyanate (MDI) [324]. Annealing of the system generates a structure with stronger hydrogen bonds.

Far infrared measurements have also been carried out as a function of temperature using FT-IR for polyethylene [318], PET [319] and polyurethanes [320]. Theory suggests that a characteristic behavior should be observed as the samples are cooled down [320]. It is expected that the peak frequency of the adsorption should increase with decreasing temperature as was observed for polyethylene and PET. The frequency shift of the maximum absorption frequency is explained by the anharmonicity of the lattice potential. The decrease of lattice distance is follwed by an increase of the force constants which explains the shift of the absorption peak to higher frequencies. For the poly-(adipine acid-glycolester) as the soft segments and the diphenylmethyl-diisocyanate as the hard segments, the bands at 96 and 174 cm^{-1} showed an increase with decreasing temperature. The band at 96 cm^{-1} is generated by pairs of benzene rings interacting in the hard or ordered phase. Isolated benzene rings in the urethanes do not contribute to the band at 96 cm^{-1}. The 174 cm^{-1} mode is assigned to the torsional vibrations of the HN—CO groups. The spectroscopic changes are consistent with the proposed structural models of the hard segment clusters of polyurethanes suggested by Bonart [321].

In most cases when the temperature of measurement is above the glass transition, the effect of temperature leads to very complicated spectral effects since structural changes and temperature-induced spectroscopic changes are occuring simulataneously [201,322]. In some cases, the structural changes are well defined as in the case of polystyrene [322].

Measurement of the absorbance changes with temperature can be used to determine the conformation energy differences in polymeric materials. It is possible to assign those bands which increase in intensity with temperature to the less-stable conformation which is expected to increase in amount with increasing temperature. One uses the convention of assigning energy by the equation

$$\Delta h_{\pm} = \frac{-R\,\partial \ln A_{\pm}(v)}{\partial(1/T)}$$

where the bands which increase in intensity are given by $A_{+}(v)$. A similar equation h_{-} can be written for those bands which decrease in intensity, i.e., A_{-}. The Van't Hoff energy is defined as

$$\Delta H = h_{+} - h_{-} = \frac{R\,\partial \ln K}{\partial(1/T)}$$

The equilibrium constant is defined by

$$K = \frac{[A_{+}(v)/\alpha_{+}]}{[A_{-}(v)/\alpha_{-}]}$$

where $A_{\pm}(v)$ and $\alpha_{\pm}$ are peak absorbances and extinction coefficients of increasing and decreasing bands, respectively. These equations assume that the extinction coefficients are independent of temperature. This assumption can be tested by plotting the ratio of A_{-} versus A_{+}. If a linear curve is obtained then the extinction coefficients are independent of temperature or have the same temperature dependence. Studies of PMMA have been reported using these equations and the results discussed briefly in section X. above.

In order to sort out the effects of temperature, it is necessary to have two different samples at the same temperature which have different structural or conformational distributions. The samples should not undergo significant structural changes during the measurement time at the temperature being investigated. Such samples can be obtained by an annealing treatment which occurs at a high enough temperature that the structural changes will have been completed and no additional structural changes will occur when the sample is held at lower temperatures for a short period of time. Figure 27 shows the effect of measurement temperature on the spectra of PET and the spectral changes are complex since conformational trans-gauche isomerization, changes in molecular packing and other modifications of the PET are occuring as the temperature is being increased. Using a highly annealed sample as a standard and a sample of PET with a uniform temperature, the absorbance ratio method can be used to determine the spectra of the trans and gauche isomers at the measurement temperatures. These spectra are shown in Fig. 28 and 29. One can now compare using absorbance subtraction these spectra at different temperatures to

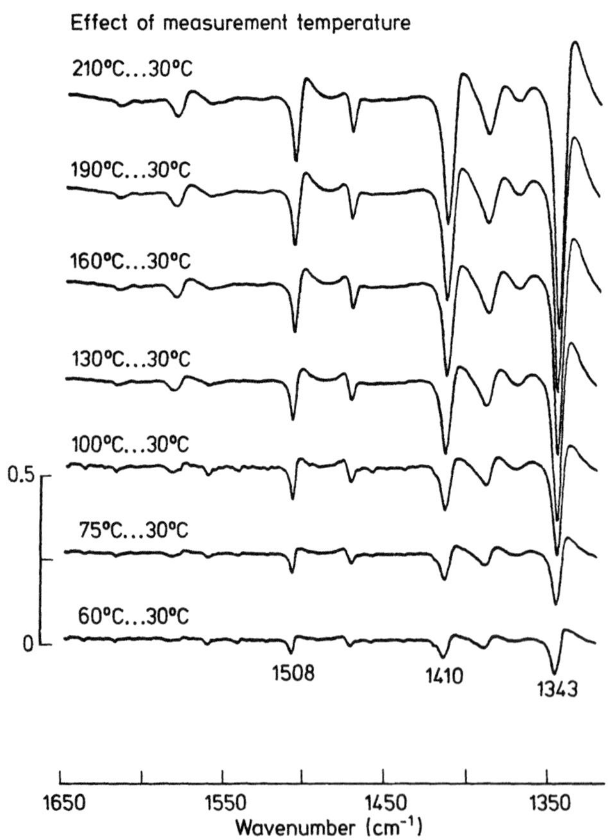

Fig. 27. Difference spectra due to reversible thermal effects only. Difference spectra of highly annealed PET film between elevated temperatures and 30 °C on the basis of one-to-one subtraction, from 1650 cm^{-1} to 1320 cm^{-1}

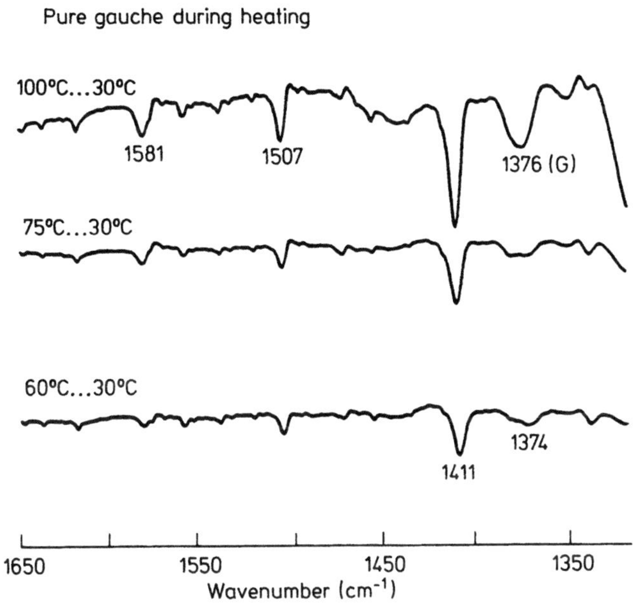

Fig. 28. Response of gauche isomer to thermal environment. Difference spectra of gauche isomers at elevated measurement temperatures, from 1650 cm^{-1} to 1320 cm^{-1}

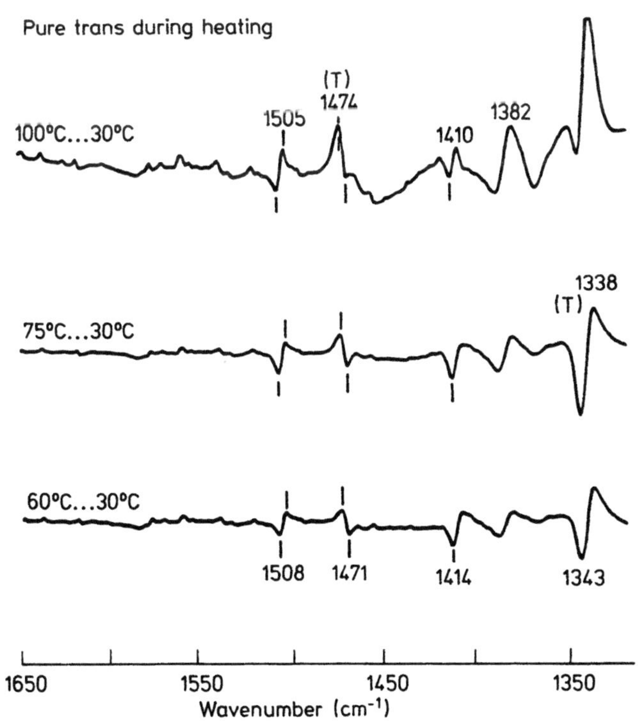

Fig. 29. Response of trans isomer to thermal environment. Difference spectra of trans isomers at elevated measurement temperatures, from 1650 cm^{-1} to 1320 cm^{-1}

determine the effect of temperature on these isomers, or one can use these spectra to calculate the trans-gauche isomeric ratio at the measurement temperature [307].

The infrared spectrum of partially deuterated polyethylene has been examined as a function of temperature [323]. Bands associated with trans-trans (tt) bond pairs and trans-gauche (tg) bonds have been identified so it is possible to follow the changes in these isomers as a function of temperature. As the temperature increased, the intensity of the tg band increased, while the intensity of the tt band decreased. It is hypothesised that the extinction coefficients for the two bands are essentially equal and the nearly constant total intensity of the two bands seems to confirm this hypothesis. Thus it appears that each time a tt sequence disappears a tg sequence is created. Correlations of this infrared result with the intensity of the Raman longitudinal acoustic mode, suggest that the new thermally induced tg sequences occur mainly as point dislocations in the crystal [323].

11 Biological Applications

Although the primary focus of this review is synthetic polymers, one would be remiss not to note the impact of FT-IR on the study of biological polymers. FT-IR has become a powerful probe of local molecular structure and environment in biological systems. Local group interactions between the biological macromolecules and their internal and external environments can be followed. Infrared studies in aqueous solutions are difficult since one must see through the strongly absorbing aqueous solution in order to observe the spectra of the biological macromolecules. However, techniques have been developed for obtaining the spectra [325] and by using difference spectroscopy the spectral changes can be detected easily [326]. Many of the applications to biological systems utilize internal reflectance spectroscopy as the sampling technique [327] because the transmission cells required to study aqueous solutions are extremely thin and require solute concentrations that are relatively high.

In spite of the obvious difficulties of the experiments and interpretation, substantial progress has been made by infrared spectroscopy in a number of areas. One of the fastest moving areas is the study of protein adsorption from blood onto various surfaces [328,329,330]. The experiments have evolved to the point where an ex vivo system is used in which a living beagle dog or a sheep provide the blood used in the protein adsorption study [330,331,341]. The initial spectra show that albumin and glycoproteins adsorb rapidly followed by fibrinogen and other proteins. Albumin is steadily replaced by other proteins until a clot is formed. The blood clotting process is being studied under conditions which will ultimately lead to new insights into this process.

Detailed insights into the molecular interactions of hemoglobin have been made using FT-IR. The molecular probe in this case was the sulfhydryl group of the alpha-104 cysteine. The center frequency shift with the strength of the hydrogen bonding and comparison of the SH frequencies in hemoglobin from horses, pigs and humans were mate [332,333].

Natural biological membranes were studied from 0 to 50 C [334]. The plasma membrane of the microorganism Acholeplasma laidlawii was compared with a model lipid (1,2-diperdeuteropalmitoyl-sn-glycero-3-phosphocholine). The phase transition of the biomembrane occurs over a 20 C range while for the model membrane the phase transition is sharp [335]. The purple membrane of Halobacterium halobium exhibits polarization properties indicating that the average orientation of the alpha-helices lies about 26 degrees away from the membrane normal. There is also spectroscopic evidence that the bacteriorhodopsin contains distorted alpha-helical conformations [336, 343, 344]. The response of Rhodopsin, the major protein component of the disc photoreceptor membranes, has been studied [337]. Delipidation leads to an alteration of the rhodopsin structure which is restored upon reconstitution.

The molecular basis of the catalytic effectiveness of enzymes remains obscure but recent studies on the nature of the substrate distortion by triosephosphate isomerase have given as new insights. Two carbonyl bands are observed for the enzyme-bound dihydroxyacetone phosphate. It appears that the difference in frequency results from the polarization of the substrate's carbonyl group by an enzymic electrophile more effective than water. This electrophile should stabilize the negative charge that accumulates on the incipient enolate oxygen in the transition state and enhance the enzyme's catalytic power [338]. The difference spectra of carbon monoxide bound to iron and copper mitochondrial cytochrome c oxidase from beef heart between light and dark were interpreted in an attempt to understand the CO inhibition by visible light [339]. A series of reactions are proposed to explain the properties of cytochrome oxidase observed with the carbon monoxide complex.

A study has been made of the DNA secondary structure induced by the various nucleohistone complexes [342]. The interaction of calf thymus DNA with the anti-cancer drug Cisplatin in water and heavy water has been studied. The carbonyl bands at 1710 and 1686 cm^{-1} of the control DNA disappear and shift to lower frequencies in the spectra of the products of the reaction. The drug induces a reorganization of the water molecules and the DNA structure is modified [340].

Acknowledgement: The author wishes to express his gratitude to the National Science Foundation for the financial support of this project under Grant Numbers, DMR80-11185 and DMR81-19425.

12 References

1. Thompson, H. W.: Proc. Roy. Soc. *A184* 21 (1945)
2. Krimm, S.: Fortschr. Hochpolym.-Forschg. *2*, 51 (1960)
3. Griffiths, P. R.: Chemical Infrared Transform Spectroscopy, Wiley, New York 1975
4. Ferraro, J. R. et al.: Fourier Transform Infrared Spectroscopy Applications to Chemical Systems, Academic Press, New York 1978, Ferraro, J. R., Basile, L. J. (Eds.)
5. Koenig, J. L.: Acct. of Chem. Res. *14*, 171 (1981)
6. Durig, J. (Ed.): Biological and Chemical Applications of FT-IR, Reidel, D., Dordrecht 1980
7. Koenig, J. L.: Appl. Spectrosc. *29*, 293 (1975)
8. D'Esposito, Koenig, J. L.: Application of Fourier Transform Infrared to Synthetic Polymers and Biological Macromolecules, in: Fourier Transform Infrared Spectroscopy, Ferraro, J. R., Basile, L. J. (Eds.) Academic Press, Vol. 1, chapter 2, 1978
9. Coleman, M. M., Painter, P. C.: J. Macro. Chem. *C16*, 197 (1978)
10. Koenig, J. L., Antoon, M. K.: Appl. Optics *17*, 1374 (1978)
11. Coleman, M. M., Painter, P. C. in: Proc. 5th Europ. Symp. Polymer Spectroscopy, Hummel, D. D. (Ed.) Verlag Chemie Weinheim 1979, p. 49
12. Siesler, H. W.: ibid., p. 137

13. Siesler, H. W.: J. Mol. Str. *59*, 15 (1980)
14. Jasse, B.: Fourier Transform Infrared Spectroscopy of Synthetic Polymers, in: Developments in Polymer Characterization, Dawkins, J. V. (Ed.) Vol. 4, p. 91, 1983
15. Painter, P. C., Coleman, M. M., Koenig, J. L.: Theory of Vibrational Spectroscopy with Application to Polymers, John Wiley and Sons, New York 1981
16. Siesler, H. W., Holland-Moritz, K.: Infrared and Raman Spectroscopy of Polymers, Marcel Dekker Inc., New York 1980
17. Michelson, A. A.: Phil. Mag. (5) *31*, 256 (1981)
18. Michelson, A. A.: ibid. *34*, 280 (1982)
19. Griffiths, P.: FT-IR. Theory and Instrumentation, in: Transform Techniques in Chemistry, Griffiths, P. (Ed.) Plenum Press 1978, p. 120
20. Stone, J. M.: Radiation and Optics, McGraw-Hill, New York 1963
21. Bell, R. J.: Introductory Fourier Transform Spectroscopy, Academic Press, New York 1982
22. Griffiths, P. R.: Chemical Infrared Fourier Transform Spectroscopy, Wiley, New York 1975, Chapter 2
23. Chamberlain, J.: Principles of Interferometric Spectroscopy, Wiley-Interscience, New York 1979
24. Champeney, D. C.: Fourier Transforms and their Physical Applications, Academic Press, New York 1973
25. Bracewell, R.: The Fourier Transform and Its Applications, McGraw-Hill, New York 1965
26. Foskett, C., in: Transform Techniques in Chemistry, Griffiths, P. E. (Ed.) Plenum Press, New York 1978 Chapter 2
27. Papoulis, A.: The Fourier Integral and Its Applications, McGraw-Hill, New York 1962
28. Mertz, L.: Transformation in Optics, John Wiley and Sons, New York 1965
29. Cooley, J. W., Tukey, J. W.: Math. Comput. *19*, 297 (1965)
30. Bloomfield, P.: Fourier Analysis of Time Series. An Introduction, John Wiley and Sons, New York 1976
31. Brigham, E. D.: The Fast Fourier Transform, Prentice-Hall, Englewood Cliffs, N.J. 1974
32. Cooper, J. W., in: Transform Techniques in Chemistry, Griffiths, P. R. (Ed.) Plenum Press 1978, p. 86
33. Jacquinot, P., Roizen-Dossier, B., in: Progress in Optics, Wold, E. (Ed.) North-Holland, Amsterdam 1964, Vol. III
34. Norton, R., Beer, R.: J. Opt. Soc. Am. *66*, 3 (1976)
35. Schau, H. C.: Infrared Physics *19*, 65 (1979)
36. Happ, H., Genzel, L.: Infrared Phys. *1*, 39 (1961)
37. Bertie, J. E., in: Analytical Applications of FT-IR to Molecular and Biological Systems, Durig, J. (Ed.) D. Reidel 1980
38. Rabolt, J., Bellar, R.: Appl. Spectrosc. *35*, 132 (1981)
39. Schroder, B., Geick, R.: Infrared Physics *18*, 595 (1978)
40. Mertz, L.: ibid. *7*, 17 (1967)
41. Forman, M. L., Steel, W. H., Vanasse, G. A.: J. Opt. Soc. Am. *56*, 59 (1966)
42. DeHaseth, J. A., Azarraga, L. V.: Anal. Chem. *53*, 2292 (1981)
43. Felegett, P. B.: J. Phy. Radium *19*, 187, 237 (1958)
44. Antoon, M. K., Koenig, J. L.: J. Polym. Sci., Polym. Chem. Ed. *19*, 549 (1981)
45. Liebman, S. A., Ahlstrom, D. H., Griffiths, P. R.: Appl. Spectrosc. *30*, 355 (1976)
46. Holland-Moritz, K., Stach, W., Holland-Moritz, I.: J. Mol. Struct. *60*, 1 (1980)
47. Holland-Moritz, K., Stach, W., Holland-Moritz, I.: Prog. Colloid and Polymer Sci. *67*, 161 (1980)
48. Jacquinot, P.: Rep. Prog. Phys. *13*, 267 (1960)
49. Griffiths, P. R., Sloane, H. J., Hannah, R. W.: Appl. Spectrosc. *31*, 485 (1977)
50. Hart, W. W., Painter, P. C., Koenig, J. L.: Appl. Spectrosc., *31*, 220 (1977)
51. Jennings, W.: Masters Thesis CWRU, Cleveland Ohio (1976)
52. Connes, J., Connes, P.: J. Opt. Soc. Am. *56*, 896 (1966)
53. Tabb, D. L., Sevick, J. J., Koenig, J. L.: J. Polym. Sci., Polym. Phys. Ed. *13*, 815 (1975)
54. Coleman, M. M. et al.: J. Polym. Sci., Polym. Lett. Ed. *12*, 577 (1974)
55. Koenig, J. L., Tabb, D. L., Coleman, M. M.: J. Polym. Sci., Polym. Phys. Ed. *13*, 1145 (1975)

56. D'Esposito, L., Koenig, J. L.: ibid. *14*, 1731 (1976)
57. Painter, P. C., Koenig, J. L.: ibid. *15*, 1885 (1977)
58. Tabb, D. L., Koenig, J. L.: Macromolecules *8*, 929 (1975)
59. Painter, P. C. et al.: J. Poly. Sci., Poly. Phys., Ed. *15*, 1223 (1977)
60. Painter, P. C. et al.: ibid. 1235 (1977)
61. Vasko, P. V., Koenig, J. L.: unpubl. results
62. Painter, P. C., Watzek, M., Koenig, J. L.: Polymer *18*, 1169 (1977)
63. Bachman, M. A. et al.: J. Appl. Phys. *50*, 6106 (1979)
64. Koenig, J. L.: Anal. Chem. *36*, 1045 (1964)
65. Hirschfeld, T., Mantz, A. W.: Anal. Chem. *51*, 495 (1979)
66. Hirschfeld, T.: Appl. Spectrosc. *30*, 552 (1976)
67. Jasse, B., Koenig, J. L.: J. Macromol. Sci. Rev., Macromol. Chem. *C17*, 61 (1979)
68. Cornell, S. W., Witenhafer, D. E., Koenig, J. L.: J. Polym. Sci. *A2*, 5, 301 (1967)
69. Schmidt, P. G.: ibid. *A1*, 1271 (1963)
70. Garton, A., Carsson, D. J., Wiles, D. M.: Appl. Spectrosc. *35*, 432 (1981)
71. Anderson, R. J., Griffith, P. R.: Anal. Chem. *47*, 2339 (1975)
72. Anderson, R. J., Griffiths, P. R.: ibid. *50*, 1804 (1978)
73. Hirschfeld, T.: Appl. Spectrosc. *30*, 549 (1976)
74. Foskett, C. T.: ibid. *30*, 531 (1976)
75. Hirschfeld, T., Kizer, K.: ibid. *29*, 345 (1975)
76. Baghdadi, A., Forman, R. A.: ibid. *35*, 473 (1981)
77. Hirschfeld, T.: Quantitative FT-IR. A Detailed Look at the Problems Involved, in: ref. 4, Ferraro, Vol. 2
78. Alben, J. O., Bare, G. H.: Appl. Optics *17*, 2985 (1978)
79. Wool, R. P.: Measurements of Infrared Frequency Shifts in Stressed Polymers, preprint
80. Koenig, J. L., Tabb, D. L.: Can. Res. and Dev. *7*, 25 (1974)
81. Koenig, J. L.: Amer. Lab. Sep., 1974, p. 9
82. Manocha, A. S. et al.: Appl. Spectrosc. *32*, 344 (1978)
83. Griffiths, P. R., Kuehl, D., Fuller, M. P.: Applications of Fourier Transform Infrared Spectrometry in Forensic Analysis, Digilab Application Note 34, April 1980
84. Battiste, D. et al.: Anal. Chem. *53*, 2232 (1981)
85. Hirschfeld, T. B.: ibid. *48*, 721 (1976)
86. Koenig, J. L., D'Esposito, L., Antoon, M. K.: Appl. Spectrosc. *31*, 292 (1977)
87. Diem, H., Krimm, S.: Appl. Spectrosc. *35* 421 (1981)
88. Koenig, J. L., Kormos, D.: Appl. Spectrosc. *33*, 351 (1979)
89. Koenig, J. L., Kormos, D.: Contemporary Topics in Polymer Science *3*, 1278 (1979)
90. Lin, S., Koenig, J. L.: J. Polym. Sci., Phys. Ed. *20*, 2277 (1982)
91. Rummel, R. J.: Applied Factor Analysis, Northwestern Univ. Press, Evanston, Ill 1970
92. Bullmer, J. T., Shurvell, H. F.: J. Phys. Chem. *77*, 256 (1900)
93. Malinowski, E. R., Lowery, D. G.: Factor Analysis in Chemistry, John Wiley, New York 1980
94. Lawley, D. N., Maxwell, A. E.: Factor Analysis as a Statistical Method, American Elsevier, New York 1971
95. Horst, P.: Factor Analysis of Data Matrices, Holt, Rinehart and Winston, New York 1965
96. Antoon, M. K., D'Esposito, Koenig, J. L.: Appl. Spectrosc. *33*, 349 (1979)
97. Malinowski, E. R.: Anal. Chem. *49*, 606 (1977)
98. Knorr, F. J., Futrell, J. H.: ibid. *51* (1979)
99. Ohta, N.: ibid. *45*, 553 (1973)
100. Tovar, M. J. M., Koenig, J. L.: Appl. Spectrosc. *35*, 543 (1981)
101. Gillette, P. C., Lando, J. B., Koenig, J. L.: Anal. Chem. *55*, 630 (1983)
102. Gillette, P. C., Lando, J. B., Koenig, J. L.: App. Spectrosc. *36*(4), 401 (1982)
103. Gillette, P. C., Lando, J. B., Koenig, J. L.: App. Spectrosc. *36*(6), 661 (1982)
104. Antoon, M. D., Zehner, B. E., Koenig, J. L.: Polym. Comp. *1*, 24 (1980)
105. Antoon, M. K., Zehner, B. E., Koenig, J. L.: ibid. *2*, 81 (1981)
106. Antoon, M. K., Koenig, J. H., Koenig, J. L.: Appl. Spectrosc. *31*, 518 (1977)
107. Haaland, D. M., Easterling, R. G.: Appl. Spectrosc. *34*, 539 (1980)
108. Haaland, D. M., Easterling, R. G.: Appl. Spectrosc. *36*, 665 (1982)

109. Gillette, P. C., Lando, J. B., Koenig, J. L.: Appl. Spectrosc. *36*, 401 (1982)
110. Gillette, P. C., Koenig, J. L.: ASTM J., in press
111. Culler, S., Ishida, H., Koenig, J. L.: FT-IR Reflectance Measurements, Ann. Rev. Mater. Sci. *13*, 363 (1983)
112. Haslam, J., Willis, H. A., Squirrell, D. C. M.: Identification and Analysis of Plastics, 2nd Ed., Heyden, Philadelphia 1972
113. Tabb, D., Koenig, J. L.: in ref. 6, p. 241
114. Reynolds, J. F. S. C.: J. Chem. Phys. *41*, 47 (1964)
115. Jakobsen, R.: J. Transform *1*, 16 (1974)
116. Lippincott, E. R. et al.: Spectrochim. Acta *16*, 58, (1960)
117. Kemeny, G., Mink, J., Laszo, F.: Kem. Kozl. *55*, 299 (1981)
118. Garton, D. J., Carlsson, Wiles, D. M.: Textile Research J. *51*, 28 (1981)
119. Tirpak, G. A., Sibilia, J. P.: J. Appl. Polym. Sci. *17*, 643 (1973)
120. Kortüm, G., Delfts, H.: Spectrochim. Acta *20*, 504 (1964)
121. Willey, R. R.: Appl. Spectrosc. *30*, 593 (1976)
122. Fuller, M. P., Griffiths, P. R.: Anal. Chem. *50*, 1906 (1978)
123. Kortüm, G.: Reflectance Spectroscopy Principles. Methods, Applications, Springer-Verlag, Heidelberg, Berlin, New York (1964)
124. Fahrenfort, J.: Spectrochim. Acta *17*, 698 (1961)
125. Fahrenfort, J., Visser, W. M.: ibid. *18*, 1103 (1962)
126. Harrick, N. J.: Phys. Rev. *125*, 1165 (1962)
127. Harrick, N. J.: Internal Reflection Spectroscopy, Wiley, New York 1967
128. Jacobsen, R.: ref. 4, chap. 5 ATR of Biological Systems
129. Paik, C. S., Hu, C. B.: Surface Chemical Analysis of Segmented Polyurethanes. Fourier Transform IR Internal Reflection Studies in "Advances in Chemistry" Series,Cooper, S., Estes, G. (Eds) *176*, 69 (1979)
130. Sung, C. S. P.: Macromol. *14*, 591 (1981)
131. Hirschfeld, T.: Appl. Spectrosc. *24*, 277 (1970)
132. Hirschfeld, T.: ibid. *32*, 160 (1978)
133. Tompkins, H. G.: ibid. *30*, 377 (1976)
134. Greenler, R. G.: J. Chem. Phys. *44*, 310 (1966)
135. Allara, D. L., Pryde, C. A.: preprint
136. Greenler, R. G.: J. Catal. *23*, 42 (1971)
137. Greenler, R. G.: J. Chem. Phys. *44*, 310 (1966)
138. Francis, S. A., Ellison, A. H.: J. Opt. Soc. Amer. *49*, 131 (1959)
139. Boerio, F. J., Schoenien, L. H., Greivenkamp, J. E.: J. Appl. Polym. Sci. *22*, 203 (1978)
140. Boerio, F. J., Chen, S. L.: Appl. Spectrosc. *33*, 121 (1979)
141. Kubelka, P., Munk, F.: Z. Tech. Phys. *12*, 593 (1931)
142. Kubelka, P.: J. Opt. Soc. Amer. *38*, 448 (1948)
143. Hecht, H. G.: J. Res. Natl. Bur. Stand., Sect. A. *80A* 567 (1976)
144. Allara, D. L., in: Characterization of Metal and Polymer Surfaces, Lee, L. H. (Ed.) Pergamon Press, New York Series IX, *11*, 1 (1972)
145. Webb, J. D. et al.: Appl. Spectrosc. *35*, 598 (1981)
146. Houghton, J., Smith, S. D.: Infrared Physics, Oxford Univ. Press, Oxford 1966
147. Harrison, T. R.: Radiation Pyrometry, John Wiley and Sons, Inc., New York 1960
148. Hadni, A.: Essentials of Modern Physics Applied to the Study of the Infrared, Pergamon, Oxford 1966
149. Kember, D. et al.: Spectrochim. Acta *35A*, 455 (1979)
150. Chase, D. B.: Appl. Spectrosc. *35*, 77 (1981)
151. Kozolowski, T. R.: Appl. Optics *7*, 795 (1968)
152. Koenig, J. L., Jennings, W.: unpubl. results
153. Low, M. J. D., Coleman, I.: Spectrochim. Acta *22*, 369 (1966)
154. Low, M. J. D., Coleman, I.: ibid. 1293 (1966)
155. Bates, J. B., Boyd, G. E.: Appl. Spectrosc. *27*, 204 (1973)
156. Griffiths, P. R.: ibid. *26*, 73 (1972)
157. Low, M. J. D.: ibid. *22*, 463 (1968)
158. Brown, R. J., Youn, B. G.: Appl. Optics *14*, 2927 (1975)

159. Bates, J. B.: in ref. 4, chapter 3
160. Kember, D., Sheppard, N.: Appl. Spectrosc. *29*, 496 (1975)
161. Sheppard, N.: The Use of Fourier Transform Methods for the Measurement of Infrared Emission Spectra, in: Analytical Applications of FT-IR to Molecular Systems, J. Durig (Ed.) D. Reidel, 1980, p. 125
162. King, V. W., Lauer, J. L.: J. Lubr. Tech. *103*, 65 (1981)
163. Viskanta, R., Hommert, P. J., Groninger, G. L.: Appl. Optics *14*, 428 (1975)
164. Steger, E., Rasmus, R.: Appl. Spectrosc. *28*, 376 (1974)
165. Baumgarten, E.: Spectrochim. Acta *A32*, 865 (1976)
166. Lauer, J. L., Peterkin, M. E.: J. Lubr. Tech. *97*, 145 (1975)
167. Lauer, J. L., Peterkin, M. E.: Amer. Lab. *7*, 27 (1975)
168. Lauer, J. L., Peterkin, M. E.: J. Lubr. Tech. *98*, 230 (1978)
169. Lauer, J. L.: ibid. *101*, 67 (1979)
170. Lauer, J. L., Peterkin, M. E.: Preprints, Div. Pet. Chem. Amer. Chem. Soc., New York Meeting, April 1976, p. 48
171. Lauer, J. L.: in ref. 4, chapter 7
172. Rosencwaig, A.: Science *181*, 697 (1973)
173. Pao, Y. H.: Optoacoustic Spectroscopy and Detection, Academic Press 1977
174. Rosencwaig, A.: Adv. in Electronics and Electron Physics, Academic Press, New York *46*, 208 (1978)
175. Rosencwaig, A.: Photoacoustics and Photoacoustic Spectroscopy, John Wiley and Sons, New York 1980
176. Busse, G., Bullemer, B.: Infrared Phys. *18*, 631 (1978)
177. Low, M. J. D., Parodi, G. A.: Spectrosc. Lett. *11*, 581 (1978)
178. Low, M. J. D., Parodi, G. A.: Infrared Phys. *20*, 333 (1980)
179. Low, M. J. D., Parodi, G. A.: Appl. Spectrosc. *34*, 76 (1980)
180. Low, M. J. D., Parodi, G. A.: J. Mol. Structure. *61*, 119 (1980)
181. Low, M. J. D., Parodi, G. A.: Spectrosc. Lett. *13*, 663 (1980)
182. Rockley, M. G.: Chem. Phys. Lett. *68*, 455 (1979)
183. Vidrine, D. W.: Appl. Spectrosc. *34*, 314 (1980)
184. Laufer, G. et al.: Appl. Phys. Lett. *37*, 617 (1980)
185. Rockley, M. G., Devlin, J. Paul: Appl. Spectrosc. *34*, 407 (1980)
186. Rockley, M. G., Davis, D. M., Richardson, H. H.: Science *210*, 918 (1980)
187. Rockley, M. G.: Chem. Phys. Lett. *75*, 370 (1980)
188. Royce, B. S. H., Teng, Y. C.: Fourier Transform Infrared Photoacoustic Spectroscopy of Condensed Phases Proceedings of the Institute of Acoustics, PAS Conference London, U.K. 1981
189. Royce, B. S. H., Teng, Y. C., Enns, J.: 1980 Ultrasonics Symp. Proc., p. 652
190. Adams, M., Kirkbright, G. F.: Analyst (London) *102*, 281 (1977)
191. Krishnan, K.: priv. comm.
192. Blank, R. E., Wakefield, T.: Anal. Chem. *51*, 50 (1979)
193. Freeman, J. J., Friedman, R. M., Reichard, H. S.: J. Phys. Chem. *84*, 315 (1980)
194. Burggraf, L. W., Leyden, D. E.: Anal. Chem. *53*, 759 (1981)
195. Lloyd, L. B. et al.: ibid. *52*, 1595 (1980)
196. Krishnan, K.: Appl. Spectrosc. *35*, 549 (1981)
197. Rosencwaig, A., Gersho, A.: J. Appl. Phys. *47*, 64 (1976)
198. Rockley, M. G.: Chem. Phys. Lett. *75*, 370 (1980)
199. Riseman, S. M. et al.: Appl. Spectrosc. *35*, 557 (1981)
200. Coleman, M. M., Petcavich, R. J., Painter, P. C.: Polymer *19*, 1243 (1978)
201. Koenig, J. L., Antoon, M. K.: J. Polym. Sci., Phys. Ed. *15*, 1379 (1977)
202. Rubcic, A., Zerbi, G.: Macromolecules *7*, 754 (1974)
203. Tabb, D. L., Koenig, J. L.: ibid. *8*, 929 (1975)
204. O'Reilly, J. M., Mosher, R. A.: J. Polym. Sci., Phys. Ed. *17*, 731 (1979)
205. O'Reilly, J. M., Mosher, R. A.: ibid. *19*, 1198 (1981)
206. O'Reilly, J. M., Mosher, R. A.: Macromolecules *14*, 602 (1981)
207. Sundararajan, P., Flory, P. J.: J. Amer. Chem. Soc. *96*, 5025 (1974)
208. Gibbs, J. H., DiMarzio, E. A.: J. Chem. Phys. *28*, 373 (1958)

209. Painter, P. C. et al.: J. Polym. Sci., Polym. Phys. Ed. *15*, 1223 (1977)
210. Painter, P. C., Watzek, M., Koenig, J. L.: Polymer *18*, 1169 (1977)
211. Painter, P. C., Koenig, J. L.: J. Polym. Sci., Polym. Phys. Ed. *15*, 1885 (1977)
212. Bachman, M. A. et al.: J. Appl. Phys. *50*, 6106 (1979)
213. Bachman, M. A., Koenig, J. L.: J. Chem. Phys. *74*, 5896 (1981)
214. D'Esposito, Koenig, J. L.: J. Polym. Sci., Polym. Ed. *14*, 1731 (1976)
215. Painter, P. C., Brozoski, B. A., Coleman, M. M.: J. Polym. Sci., Phys. Ed. *20*, 1069 (1982)
216. Dirlikov, S., Koenig, J. L.: Appl. Specrosc. *33*, 555 (1979)
217. Dirlikov, S., Koenig, J. L.: Appl. Spectrosc., *33*, 551 (1979)
218. Kapur, R. S., Koenig, J. L., Shelton, J. R.: Rubber Chem. *47*, 911 (1974)
219. Tabb, D. L., Sevcik, J. J., Koenig, J. L.: J. Polym. Sci., Phys. Ed. *13*, 815 (1975)
220. Pecsok, R. L. et al.: Rubber Chem. and Tech. *49*, 1010 (1976)
221. D'Esposito, L., Koenig, J. L.: Polym. Sci. and Eng. *19*, 162 (1979)
222. Shelton, J. R., Pecsok, R. L., Koenig, J. L.: Fourier Transform Infrared Studies of the Uninhibited Autoxidation of Elastomers, in: Durability of Macromolecular Materials, ACS Symp. Series 95, Eby, R. K. (Ed.) 1979, p. 75
223. Pecsok, R. L., Shelton, J. R., Koenig, J. L.: Polym. Degrad. and Stab. *3*, 161 (1981)
224. Petcavich, R. J., Painter, P. C., Coleman, M. M.: Polymer *19*, 1249 (1978)
225. Coleman, M. M., Petcavich, R. J.: J. Polym. Sci., Polym. Phys. Ed. *16*, 821 (1978)
226. Petcavich, R. J., Painter, P. C., Coleman, M. M.: ibid. *17*, 165 (1979)
227. Coleman, M. M., Sivy, G. T.: Fourier Transform Infrared Studies of the Degration of Poly-(acrylonitrile) Copolymers I. Introduction and Comparative Rates of the Degradation of Three Copolymers Below 200 °C and Under Reduced Pressure. Preprint submitted to CARBON
228. Coleman, M. M. et al.: Studies of the Degradation of Acrylonitrile/Acrylamide Copolymers as a Function of Composition and Temperature. Preprint submitted to CARBON
229. Varnell, D. F., Runt, J. P., Coleman, M. M.: FT-IR and Thermal Analysis Studies of Blends of Poly(Caprolactone) and Homo- and Copolymers of Poly(vinylidene chloride). Preprint submitted to CARBON
230. Coleman, M. M.: pers. comm.
231. Lin, S. C., Bulkin, B. J., Pearce, E. M.: J. Polym. Sci., Polym. Phys. Ed. *17*, 3121 (1979)
232. Antoon, M. K., Koenig, J. L.: ibid. *19*, 197 (1981)
233. Antoon, M. K., Koenig, J. L., Serafini, T.: ibid. *19*, 1567 (1981)
234. Antoon, M. K., Koenig, J. L.: Macromol. Sci. Rev. *C19*, 153 (1980)
235. Vanni, H., Rabolt, J. F.: J. Polym. Sci., Polym. Phys. Ed. *18*, 587 (1980)
236. DeVries, K. L., Smith, R. H., Fanconi, B. M.: Polymer *21*, 949 (1980)
237. Tabb, D., Koenig, J. L.: unpubl. results
238. Antoon, M. K., Starky, K. M., Koenig, J. L., in: Composite Materials. Testing and Design, Tsai, S. W. (Ed.) ASTM Philadelphia, PA, 1979, p. 541
239. Antoon, M. K., Zehner, B. E., Koenig, J. L.: Polym. Composites *1*, 24 (1980)
240. Hartshorn, J. H.: Division of Org. Coat. and Plastics. ACS preprints, *40*, 500 (1979)
241. Roush, P. B.: ibid. *41*, 606 (1980)
242. Chen, C. H. et al.: Rubber Chem. and Tech. *54*, 734 (1981)
243. Chen, C. H. et al.: ibid. *55*, 103 (1982)
244. Chen, C. H. et al.: ibid. 4, 1221 (1982)
245. D'Esposito, L., Koenig, J. L.: Polym. Sci. and Eng. *19*, 162 (1979)
246. Jasse, B., Koenig, J. L.: Polymer *22*, 1040 (1981)
247. Coleman, M. M. et al.: J. Polym. Sci., Polym. Lett. Ed. *15*, 745 (1977)
248. Wellinghoff, S. T., Koenig, J. L., Baer, E.: J. Polym. Sci., Polym. Phys. Ed. *15*, 1913 (1977)
249. Allara, D. L.: Appl. Spectrosc. *33*, 358 (1979)
250. Allara, D. L., Baca, A., Pryde, C. A.: Macromolecules *11*, 1215 (1978)
251. Naito, K. et al.: ibid. *11*, 1260 (1978)
252. Coleman, M. M., Zarian, J.: J. Polym. Sci., Phys. Ed. *17*, 837 (1979)
253. Coleman, M. M., Varnell, D. F.: ibid. *18*, 1403 (1980)
254. Coleman, M. M., Varnell, D. F., Runt, J. P., in: Contemporary Topics in Polymer Science, Bailey, W. J. (Ed.) Vol. 4, 1982
255. Varness, D. F., Runt, J. P., Coleman, M. M.: Macromolecules *14*, 1350 (1981)

256. Lefebvre, D., Jasse, B., Monnerie, L.: Fourier Transform Infrared Study of Unaxially Oriented Poly(2,6-dimethyl 1,4-phenylene oxide)-Atactic Polystyrene Blends, submitted to Polymer
257. Ishida, H., Koenig, J. L.: J. Coll. Interface Sci. *64*, 555 (1978)
258. Ishida, H., Koenig, J. L.: ibid. *64*, 565 (1978)
259. Ishida, H., Koenig, J. L.: J. Polym. Sci., Polym. Phys. Ed. *17*, 1807 (1979)
260. Ishida, H., Koenig, J. L.: ibid. *17*, 574 (1979)
261. Ishida, H., Koenig, J. L.: ibid. *18*, 233 (1980)
262. Ishida, H., Koenig, J. L.: ibid. *18*, 1931 (1980)
263. Ishida, H., Koenig, J. L., in: Silylated Surfaces, Midland Macromolecular Monographs No. 7, Lyden, D. E., Collins, W. (Eds.) Gordon and Breach, New York 1980, p. 73
264. Chiang, C. H., Ishida, H., Koenig, J. L.: J. Coll. Interface Sci. *74*, 396 (1980)
265. Chiang, C. H., Koenig, J. L.: ibid. *83*, 361 (1981)
266. Chiang, C. H., Koenig, J. L.: Polym. Composite *1*, 88 (1980)
267. Chiang, C. H., Koenig, J. L.: J. Coll. Interface Sci. *83*, 2, 361 (1981)
268. Chiang, C. H., Koenig, J. L.: Polym. Composite *2*, 192 (1981)
269. Ishida, H., Chiang, C. H., Koenig, J. L.: Polymer *23*, 251 (1982)
270. Ishida, H. et al.: J. Polym. Sci., Polym. Phys. Ed. *20*, 701 (1982)
271. Emadipour, H., Chiang, C. H., Koenig, J. L.: Mechanica *5*, 165 (1982)
272. Ishida, H., Naviro, S., Koenig, J. L.: J. Polym. Sci., Polym. Phys., Ed. *20*, 701 (1982)
273. Chiang, C. H., Koenig, J. L.: ibid. 2135 (1982)
274. Ishida, H., Koenig, J. L., Gardner, K. C.: J. Chem. Phys. *77*, 5748 (1982)
275. Ishida, H., Naviroj, S., Koenig, J. L.: The influence of a Substrate on the Surface Characteristics of Silane Layers, in: Physicochemical Aspects of Polymer Surfaces, Mittal, K. L. (Ed.) Plenum Press, New York 1982
276. Koenig, J. L., Chiang, C. H.: In Situ Analysis of the Interface, Proc. 1981 CAPRI Conf. Plenum Press, New York 1982
277. Chiang, C. H., Liu, N. I., Koenig, J. L.: J. Coll. Interface Sci. *86*, 26 (1982)
278. Xue, G., Koenig, J. L.: The Chemical Reactions of an Epoxy-Functional Silane in Aqueous Solution, paper presented 1983 Ahesion Society Meeting
279. Xue, G. et al.: The Reinforcement Mechanism of Polyester Fiber-Reinforced Rubber, J. Appl. Polym. Sci., submitted
280. Wool, R. P.: Polym. Eng. and Sci. *20*, 807 (1980)
281. Zhurkov, S. N. et al.: Proc. 2nd Internat. Conf. Fracture, Brighton, U.K. 1969
282. Sikka, S. S., Kausch, H. H.: Coll. and Polym. Sci. *257*, 1060 (1979)
283. Ishida, H. et al.: Macromolecules *13*, 826 (1980)
284. Wellinghoff, S. T. et al.: ibid. *13*, 834 (1980)
285. Shen, D. Y., Hsu, S. L.: Vibrational Spectroscopic Characterization of Rigid Rod Polymers. III. Microstructural Changes in Stressed Polymers, Polymer
286. Garton, A. et al.: J. Appl. Polym. Sci. *25*, 1505 (1980)
287. Jasse, B., Koenig, J. L.: J. Polym. Sci., Polym. Phys. Ed. *17*, 799 (1979)
288. Garton, A., Carlsson, D. J., Wiles, D. M.: Appl. Spectrosc. *35*, 432 (1981)
289. Krishnan, K.: ibid. *32*, 549 (1978)
290. Leonhard, C. et al.: FT-IR Evidence of Beta Crystal Phase Formation in PVDF/PMMA Blends, Polymer, in press
291. Siesler, H. W.: Polymer Deformation A Rheo-Optical Study by Fourier Transform Infrared Spectroscopy, preprint
292. Siesler, H.: Makromol. Chem. *180*, 2261 (1979)
293. Lephardt, J. O., Vilcins, G.: Appl. Spectrosc. *29*, 1140 (1973)
294. Stambaugh, B., Koenig, J. L., Lando, J. B.: J. Polym. Sci., Polym. Phys., Ed. *17*, 1063 (1979)
295. Holland-Moritz, K., van Werden, K.: Macromol. Chem. *182*, 651 (1981)
296. Holland-Moritz, K., Holland-Moritz, I., van Werden, K.: Coll. and Polym. Sci. *259*, 156 (1981)
297. Bayer, G., Hoffman, W., Siesler, H. W.: Polymer *21*, 235 (1980)
298. Stambaugh, B., Lando, J. B.: Polym. Sci., Polym. Phys. Ed. *17*, 1063 (1979)
299. Gillette, P. C. et al.: Polymer *23*, 1759 (1982)
300. Murphy, R. E., Cook, F. H., Sakai, H.: J. Opt. Soc. Amer. *65*, 600 (1975)

301. Sakai, H., Murphy, R. E.: Appl. Optics *17*, 1342 (1978)
302. Honigs, D. E. et al.: Time Resolved Infrared Interferometry I, in: Vibrational Spectra and Structure, Vol. 11, 219. Durig, J. R. (Ed.) Elsevier, N.Y. 1982
303. Garrison, A. A. et al.: Appl. Spectrosc. *34*, 399 (1980)
304. Fateley, W. G., Koenig, J. L.: J. Polym. Sci., Polym. Lett. Ed. *20*, 445 (1982)
305. Fateley, W. G., Koenig, J. L.: Time Resolved Spectroscopy of Stretched Polypropylene Films, in: Recent Advances in Analytical Spectroscopy, p. 291, Fewa, Keiichiro (Ed.) Pergamon Press, Oxford 1982
306. Burchell, D. J. et al.: Deformation Studies of Polymers by the Time Resolved Fourier Transform Infrared Spectroscopy I. Development of the Technique, Polymer, in press
307. Lin, S. B., Koenig, J. L.: J. Polym. Sci., Phys. Ed., *20*, 2277 (1982)
308. Ovander, L. N.: Opt. Spectrosc. (USSR) *11*, 68 (1961)
309. Hannon, M. J., Koenig, J. L.: J. Polym. Sci. *A-2* 7 (1969)
310. Huang, Y. S., Koenig, J. L.: J. Appl. Polym. Sci. *15*, 1237 (1971)
311. Hart, W. W., Koenig, J. L., in: Probing Polymers, Koenig, J. L. (Ed.) ACS 1979
312. Enns, J. B. et al.: Polym. Eng. Sci. *19*, 756 (1979)
313. Anton, A.: J. Appl. Polym. Sci. *12*, 2117 (1968)
314. Bessler, V. E., Bier, G.: Makromol. Chem. *122*, 30 (1969)
315. Schroeder, L. R., Cooper, S. L.: J. Appl. Phys. *47*, 4310 (1976)
316. Ogura, K., Sobue, H., Nakamura, S.: J. Polym. Sci., Polym. Phys. Ed. *11*, 2079 (1973)
317. Senich, G. A., MacKnight, W. J.: Macromol. *13*, 106 (1980)
318. Frank, W., Schmidt, H., Wuff, W.: J. Polym. Sci., Polymer Symp. *61*, 317 (1977)
319. Frank, W., Schmidt, H.: priv. comm.
320. Frank, W., Strohmeier, W.: Prog. Colloid and Polym. Sci. *66*, 205 (1979)
321. Bonart, R., Morbitzer, L., Muller, E. H.: J. Macromol. Sci., Polym. Phys. *B9*, 447 (1974)
322. Jasse, B., Bokobza, L.: J. Mol. Struct. *73*, 1 (1981)
323. Reneker, D. H., Mazur, J.: J. Appl. Phys. *51*, 5080 (1980)
324. Brunette, C. M., Hsu, S. L., MacKnight, W. J.: Macromol. *15*, 71 (1982)
325. Low, M. J. D., Yang, R. T.: Spectrochimica Acta *29A*, 1761 (1978)
326. Tabb, D. L., Koenig, J. L.: Infrared Spectra of Globular Proteins in Aqueous Solution, in: Analytical Applications of FT-IR to Molecular and Biological Systems, Durig, J. R. (Ed.) D. Reidel 1980, p. 241
327. Gendreau, R. M., Jakobsen, R. J.: Appl. Spectrosc. *32*, 326 (1978)
328. Jacobsen, R. J., Gendreau, R. M.: Antific. Organs *2*, 183 (1978)
329. Gendreau, R. M., Jakobsen, R. J.: J. Biomed. Mat. Res. *13*, 893 (1979)
330. Gendreau, R. M. et al.: Appl. Spectrosc. *35*, 353 (1981)
331. Alben, J. O., Bare, G. H., Bromberg, P. A.: Nature *252*, 736 (1974)
332. Bare, G. H., Alben, J. D., Bromberg, P. A.: Biochemistry *14*, 1578 (1975)
333. Alben, J. O., Bare, G. H.: Appl. Optics *17*, 2985 (1978)
334. Casal, H. L. et al.: Biochimica et Biophysica Acta *550*, 145 (1979)
335. Mantsch, H. et al.: Mol. Struct. *60*, 263 (1980)
336. Rothschild, K. J., Clark, N. A.: Biophys. J. *25*, 473 (1979)
337. Rothschild, K. J., DeGrip, W. J., Sanches, R.: Biochimica et Biophysica Acta *596*, 338 (1980)
338. Belasco, J. G., Knowles, J. R.: Biochem. *19*, 472 (1980)
339. Alben, J. O. et al.: Proc. Natl. Acad. Sci. Biochem. *78*, 234 (1981)
340. Theophanides, T.: Appl. Spectrosc. *35*, 461 (1981)
341. Winters, S. et al.: ibid. *36*, 404 (1982)
342. Liquier, M. C. et al.: Nucleic Acids Res. *6*, 1479 (1979)
343. Rothschild, K. J., Zagaeski, M., Cantore, W. A.: Biochem. and Biophys. Res. Comm. *103*, 483 (1981)
344. Rothschild, K. J., Marrero, H.: Proc. Natl. Acad. Sci. USA Biophys. *79*, 4045 (1982)

Received April 15, 1983
M. Gordon (editor)

Author Index Volumes 1–54

Subject Index

Key words set in *italics* refer to the headings.

GPSR Compliance

The European Union's (EU) General Product Safety Regulation (GPSR) is a set of rules that requires consumer products to be safe and our obligations to ensure this.

If you have any concerns about our products, you can contact us on ProductSafety@springernature.com

In case Publisher is established outside the EU, the EU authorized representative is:

Springer Nature Customer Service Center GmbH
Europaplatz 3
69115 Heidelberg, Germany

Batch number: 10620881

Printed by Printforce, the Netherlands